SLOPOVERS

To the Last Smoke

SERIES BY STEPHEN J. PYNE

STEPHEN J. PYNE

SLOPOVERS

Fire Surveys of the Mid-American
Oak Woodlands, Pacific Northwest, and Alaska

THE UNIVERSITY OF
ARIZONA PRESS
TUCSON

The University of Arizona Press
www.uapress.arizona.edu

© 2019 by The Arizona Board of Regents
All rights reserved. Published 2019

ISBN-13: 978-0-8165-3879-9 (paper)

Cover design by Leigh McDonald
Cover photo by Philip Spor, courtesy of Alaska Department of Forestry

Library of Congress Cataloging-in-Publication Data are available at the Library of Congress.

Printed in the United States of America
♾ This paper meets the requirements of ANSI/NISO Z39.48-1992 (Permanence of Paper).

To Sonja
old flame, eternal flame

৵

CONTENTS

SERIES PREFACE

To the Last Smoke

WHEN I DETERMINED to write the fire history of America in recent times, I conceived the project in two voices. One was the narrative voice of a play-by-play announcer. *Between Two Fires: A Fire History of Contemporary America* would relate what happened, when, where, and to and by whom. Because of its scope it pivoted around ideas and institutions, and its major characters were fires or fire seasons. It viewed the American fire scene from the perspective of a surveillance satellite.

The other voice was that of a color commentator. I called it *To the Last Smoke*, and it would poke around in the pixels and polygons of particular practices, places, and persons. My original belief was that it would assume the form of an anthology of essays and would match the narrative play-by-play in bulk. But that didn't happen. Instead the essays proliferated and began to self-organize by regions.

I began with the major hearths of American fire, where a fire culture gave a distinctive hue to fire practices. That pointed to Florida, California, and the Northern Rockies, and to that oft-overlooked hearth around the Flint Hills of the Great Plains. I added the Southwest because that was the region I knew best. The Interior West beckoned because I thought I knew its central theme and wanted to learn more about its margins. Then there were stray essays on places and themes that needed to be corralled into a volume, and there were all those relevant regions that needed at least token treatment. Some like the Lake States and Northeast no longer commanded the national scene as they once had, but their stories were interesting and needed recording, or like the Pacific Northwest or central oak woodlands spoke to the evolution of fire's American century in a new way. Alaska boasts its own regional subculture. I would include as many as possible into a grand suite of short books.

My original title now referred to that suite, not to a single volume, but I kept it because it seemed appropriate and because it resonated with my own relationship to fire. I began my career as a smokechaser on the North Rim of Grand Canyon in 1967. That was the last year the National Park Service hewed to the 10 a.m. policy and we rookies were enjoined to stay with every fire until "the last smoke" was out. By the time the series appears, 50 years will have passed since that inaugural summer. I no longer fight fire; I long ago traded in my pulaski for a pencil. But I have continued to engage it with mind and heart, and this unique survey of regional pyrogeography is my way of staying with it to the end.

Some funding for the project came from the U.S. Forest Service, Department of the Interior, and Joint Fire Science Program, part of the residual monies left after researching *Between Two Fires* and the early volumes of *To the Last Smoke*. The Bureau of Land Management and Joint Fire Science Program contributed a supplement to allow me to complete the suite. I'm grateful for their support. Thanks to Kerry Smith for once again saving me from my worst grammatical self. And of course the University of Arizona Press deserves praise as well as thanks for seeing the resulting texts into print.

PREFACE TO VOLUME 8

BEN FRANKLIN ONCE MARVELED how "convenient a thing it is to be a reasonable creature, since it enables one to find or make a reason for everything one has a mind to do." Since *Slopovers* is a departure in design and style from the other volumes in *To the Last Smoke*, I feel the need to rationalize a little myself. Why compress three regional surveys into one book?

The thematic reasons are that these regions have not contributed directly to the fire revolution that was the subject of *Between Two Fires* and do not have fire cultures that affect the national scene on the scale of the other regions in the series. They were historically important, even commanding, at one time, but less so now. They have not shifted their significance since what I described in *Fire in America*. The practical reasons are money and time. I could do short surveys with the residual funds and time left from my original grant. I haven't the calendar space to give every region the attention each believes—rightly—it deserves.

Slopovers is a compromise, going beyond the borders of my founding conception, but not much over. The spillage is contained (I trust). My experiment in writing a minisurvey on Texas for the Great Plains volume showed the literary possibilities of what might be called a nonfiction novella—a structural analogy, not a statement that the material is in the least made up. I couldn't bear to leave the regions out of the larger suite, yet I didn't have the resources to commit to full-bore surveys.

The sequencing of regions broadly conforms to the time when each best revealed the national fire saga. The oak woodlands and savannas were most active in the era of frontiering across the Appalachians and into the middle border, roughly the late 18th century and early 19th. The Pacific Northwest picked up and carried the torch from the early to mid-20th century. Alaska only acquired the apparatus of wildland fire management in 1939; its seminal contribution, the Alaska Fire Service and Alaska Interagency Fire Management Plan, arose during the early 1980s. Still, my *Slopover* has itself a slopover in the Klamath Mountains. I had run out of space in my California survey (to the dismay of reviewers for the press), so I elected to add them to the Northwest, with the result that I have only run out of space here as well. Maybe some places just need to be themselves.

Today, all three regions occupy special niches in America's pyrogeography. If their operations somehow shut down, it's difficult to imagine the major thrust of American fire changing much. Yet the regions matter. They display an important diversity of fire, fire ideas, and fire practices, helping to check national ambitions that might be inappropriate across the map; they occupy large tracts of the national estate; and they may push themselves back into the national limelight at some future time, as the Pacific Northwest seems to be doing in recent years. They are interesting in their own right, and if they occupy secondary tiers, they are closer to the top tier than to the lower. They matter. It's been fascinating to learn more of their recent evolutions.

Those who made my visit productive (and in some cases, possible) are acknowledged in the individual essays. But I offer my collective thanks again here, as well as to the University of Arizona Press for allowing me a slopover in my word count.

SLOPOVERS

THE MID-AMERICAN OAK WOODLANDS

A FIRE SURVEY

"Now called Kentucke, but known to the Indians by the name of the Dark and Bloody Ground, and sometimes as the Middle Ground."
—JOHN FILSON, *THE DISCOVERY, SETTLEMENT AND PRESENT STATE OF KENTUCKE* (1784)

The word Kentucky means . . .
"The Prairie, or Barrens" (Catawba)
"among the meadows" (Mohawk)
"Place of the Meadows" (Delaware)[1]

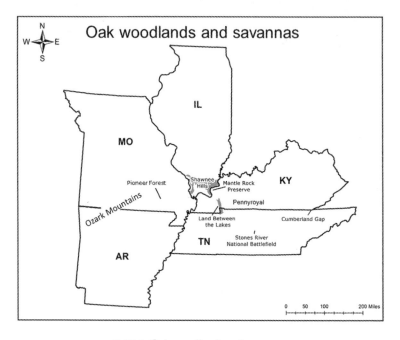

MAP 1 Oak woodlands and savannas.

AUTHOR'S NOTE

Oak Woodlands

I GOT INTRODUCED TO the region at a workshop in 2014, then organized a fire study tour in 2015. It then took a long time to create the publication venue a minisurvey of this kind requires. The months passed, then years. Yet my memory remains vivid. For that I can thank the many colleagues who donated their time and knowledge to educate me into a fire region I knew little about. The surprise, for me, was how much (of a sort) I did know regarding the region from previous studies in geomorphology and the American frontier, and my long interest in Carl Sauer. It was fun to overlay those pieces and map out a strategy for a brief fire survey.

In this text *oak* is shorthand for *oak-hickory*, and *woodlands*, for the barrens, woodlands, and savannas that characterize the biota.

PROLOGUE

East of the 100th Meridian

I N 1893 FREDERICK JACKSON TURNER read one of the most famous sentences ever penned by an American historian. "Stand at Cumberland Gap and watch the procession of civilization, marching single file—the buffalo following the trail to the salt springs, the Indian, the fur-trader and hunter, the cattle raiser, the pioneer farmer—and the frontier has passed by." A colonial society, hemmed in between the Atlantic Ocean and the Appalachian Mountains, suddenly spilled through the gap and then seized a continent. Within a single long generation, a folk migration blew across the Appalachians to the Great Plains, from mixed woodlands to prairie. The rush of settlement poured through the gap, and simultaneously down the Ohio River, like a stream debouching into a long delta before slowing and spreading to its flanks. This saga is, more than any other, America's favorite creation story.[1]

Less well appreciated, that historical process overlays an ecological biome that was virtually coextensive with it. The American backwoods frontier was primarily a backwoods of oak, hickory, and grass. The sprawl of pioneering filled the oak woodlands and the prairie barrens like water in a basin before splashing over north and south. The early settlers didn't linger in the mountains: they hurried through them and across the deeply dissected Cumberland Plateau where it joins the Appalachians until the plateau lowered into hills and they reached the oak-hickory savannas and great barrens, dense with game and grass, and there they settled. It was a biome peculiarly suited to their simple economies of hunting, trapping, herding, and resettling. Like their archetypal guide, Daniel Boone, they moved in and then moved on, and then moved on again, until the hills flattened into the plains and woodland mosaics thinned into tallgrass prairie.

Until recently neither narrative—the saga of pioneering or the tale of a sprawling biome that spanned central Middle America—spoke to fire. There is a long tradition of discourse over the origin of the grassy Barrens, which splashed about the region and became the rallying points for American settlement. On one side stood those who argued for the unquestioned supremacy of climate: the prairies were barren of trees because climate and soil (itself a byproduct of climate) dictated that only forbs and grasses and scattered shrubs could grow there. On the other were those who promoted fire, which was also a declaration on behalf of the power of humans to challenge climate since natural fire was almost unknown. A related, secondary skirmish broke out over the role of fire along the fringe where oak and prairie met. But fire in the hardwoods themselves was dismissed. It was unthinkable that fire might have any integral role; it appeared only as an unwanted disturbance or act of ecological vandalism. On this nearly all authorities agreed.

Yet the evidence builds that the great oak-hickory forests were fire frequented and probably fire informed. And because natural fire has almost no presence, this means those obligatory fires were set by humans, and had to have been set by people for millennia, perhaps as far back as the origin of the Hypsithermal, and this realization changes the narrative of pioneering. If correct it means that America's indigenous peoples, through fire, had been instrumental in sculpting the great swath of woodlands and grasslands that swept through the middle American lands east of the Mississippi. America's backwoods frontiersmen picked up that torch. Instead of something incidental to the scene, like a campfire, outside the major action, the realization grows that anthropogenic fire is a critical catalyst, a core technology that helped make the rest possible. Remove it, and the landscape unravels. That, too, has happened. And it threatens to redefine, if not unhinge, the intellectual geography of fire in America.

In the West it was possible to pretend that pre-Columbian humans had little say in the grand processes that shaped landscapes. In the oak woodlands that is not possible, even in the imagination. Here natural history and human history flow in and out of each other, like streams in the karstified Barrens, that rise and fall, flow and disappear, with seasons and sinkholes. Human history, fire history—neither makes sense without the other.

THE LONG HUNT

AN APOCRYPHAL STORY of the young Daniel Boone has him "fire hunting," which in the phraseology of the time, meant hunting at night while a companion held a torch that would attract deer and cause their eyes to shine and hence mark them as targets. Instead, the flames caught the eyes of Rebecca Bryan looking for stray cattle. Boone held his fire while the reflected fire in Rebecca's eyes enchanted him. Within months they were married. According to the legend, Boone renounced further fire hunting.[1]

Widen the literary lens from folktale to allegory, however, and the story speaks to the encounter of questing Long Hunters with the Edenic, game-abounding "Kentucke." They carried fire as their Indian predecessors had, but once they found their promised land, they stayed their hand, or more accurately, they changed how they carried fire and to what ends. The practice of "fire hunting" by torchlight (what today is known as spotlighting) was so effective it was banned in the Atlantic colonies. But there were other forms of hunting by fire that involved drives and surrounds and that lured animals to hunting grounds by timed burns and green ups. Those practices endured—had to endure or the prime game, all of which fed on postburn regrowth, would go elsewhere, or would abandon sites as they overgrew to woods, and the hunters would have to follow them.

Widen that lens a bit further, and you have the story of the land itself and the hunt for an explanation for why a nominal woodlands—a land certainly capable of bearing an omnium gatherum of trees—held so much grass and over vast swathes yielded to prairie. The reason was certainly known to those who hunted on it. An explanation, however, was not apparent to those intellectuals, naturalists, foresters, and professors who pondered the question from afar. The long hunt for a cause has continued into the present day.

Aldo Leopold once observed that five tools had changed the land since contact—the gun, the torch, the cow, the ax, and the plow. That was certainly true for the trans-Appalachian frontier, and in probably that order of importance. Additionally, what made pioneer penetration possible were the diseases and war-galvanized disruptions that caused a demographic collapse among the indigenous peoples and left the remnants of tribes to gather as best they could to resist the flow of settlers that streamed into Kentucky. The Indian wars were brutal—savage on both sides. But the numbers killed were small (the famous siege of Boonesborough in 1777 involved a couple hundred Shawnee and a few score settlers).

The fact is, at the time of American expansion few indigenes resided permanently in the "Dark and Bloody Ground" known as Kentucky, or more specifically those prairies in the Barrens. Human history dated back to the waning Pleistocene, but populations and livelihoods had waxed and waned with climate, migrations, plagues, and the arrival of cultigens. Agriculture meant shifting cultivation, a fire-fallow farming best expressed in the bottomlands, and here villages clustered. The long-fallow fields would be rich in shrubs, berries, and fruits, perhaps pruned by selective fire, while the drier uplands, kept open or as savanna by regular burning, flourished with game—the largest (deer, elk, and bear) made up 90 percent of dietary meat. The wetter landscape was likely a mosaic of worked and abandoned patches. The Barrens and outlying fields were dappled with copses and glades, locally dense with canebrake and spacious with savannas. Like their outlying fields, the towns moved, twisting the landscape kaleidoscope. The Fort Ancient culture, best expressed by settlements like that at Cahokia, crested between 1000 and 1500 CE, then declined, first slowly, then precipitously. By 1550 permanent settlement in the greater Barrens had yielded to seasonal use as a hunting ground open to many peoples. It's likely that those fabled grasslands expanded after the settled Cahokia people abandoned the region, and fire, bison, and passing hunters moved into the vacuum. Bison shifted from browse to warm-season (C_4) grasses. Together they kept the land as prairie. It remained so for the next 200 years.[2]

By the mid-18th century the Barrens were indeed contested land. Diseases had hollowed out former populations, settled people relocated north of the Ohio River, tribes fractured by European contact farther east were moving into the region, and American colonists filed through Appalachian passes and down the Ohio. The land was far from empty of people, but it was seasonally vacant, as groups passed through to hunt, typically on the Warriors Path, north and south, fighting over access to those same lush sites. The American pioneers were able to establish their beachheads largely because they moved into a land further depopulated by disease and dislocation. It was less a virgin land than a widowed one. "Dark and Bloody" is apparently a mistranslation, but one that seemed apt since the land was dark with clouds from constant fighting, and it was bloody for the same reason.[3]

Later, as hunting cabins matured into farms, the ax made inroads into timber, locally quickened by the needs of salt boilers and iron furnaces, and the plow spread from patchy fields across valleys and over uplands; and to them we could add the shovel, as a means of draining wetlands. Add, too, the introduction of aliens, especially weeds. The prime Kentucky settlements nestled into what was known then as the Barrens but became known as the Bluegrass region because the trekkers unwittingly brought with them European cool-season grasses, notably bluegrass and timothy, that were better suited to the domesticated livestock that had accompanied them across the Atlantic. The Barrens, clothed in warm-season grasses and forbs, not unlike those that stocked tallgrass prairies, had seasonally dried and burned. The Bluegrass belt greened during the early spring before those traditional fires could carry.

The pioneers had lived off the hunt, supplemented by some grain, milk, and meat from cattle and hogs. In short order, they destroyed the conditions that had made that life possible. The game was gone, and not simply the creatures themselves but the habitats that had sustained them. Notoriously, those firstcomers moved on, as later generations of westering fur trappers, row-crop swiddeners, placer miners, big-tree loggers, and cattle barons would. The folklore has it that the long hunters felt crowded when they could see the smoke from a neighbor's chimney. What they meant was, they could no longer live by the long hunt as they could before. The Barrens of the contact era flourished in game because no one resided there permanently: people came only seasonally. Now, they lived in fixed plots and quickly exhausted the local game. If they wanted to hunt after settlement, they had to move on. Some woodlands had been cleared (a few areas intensively for charcoal and wood fuel to run salt works and iron furnaces), but other woods had sprung up; saplings replaced old growth. Behind them, they left savannas and barrens choking on exotic grasses and fast silting in with new trees since intensive grazing by domestic livestock chewed up the grasses that had carried flame and that change in burning had allowed young forests to seize the land like weeds.

It was not the smoke from the neighbor that mattered, but the lost smoke from fires that had free-ranged with the bison and turkey. Instead, a different regimen of fire established itself. There were fires to clear lands, to reclaim fields from fallow, to spark rough pasture and habitat from upland woods, and fires that expanded from mobile campfire to stone hearth, blacksmith's forge, and iron-smelting furnace. The torch passed from Shawnee and Cherokee to long hunter and pioneer, and then to small farmer and open-range herder. The fabled Barrens evolved into permanent pastures and arable lands. Cut and recut, the hardwoods declined. Poor plowing on hills led to soil erosion. Firing continued as a folk practice, but in reality "woodsburning" meant the burning of the oak leaves and grasses that were the surface matrix of the oak-hickory forest. Unburned land was unusable land, but as livestock differed from wildlife, so these new fires behaved differently than those that had preceded them. Still, folk burning was endemic. In Charles Sargent's forest survey for the 1880 census, only five states burned more land than Kentucky.[4]

———

This grand saga has plenty of subplots and sidebars, and that extraordinary surge of settlement was followed by a slower infilling that took decades to break the land into counties, estates, paddocks, arable fields, and towns. In their early decades Kentucky and Tennessee contributed politicians of national stature. But then the bustle of settlement moved north and south, and transcontinental traffic routed around America's middle east. Subsequent American history seemed to pass the region by, and neither the nation nor the region seemed to care. Even state and regional historians typically ended their chronicle with the Civil War and Reconstruction. The national narrative moved west and into cities dense with commerce and industry.

What had been a main current of pioneering and national politics slowed into a backwater, better known for poverty, tobacco, and a hemorrhaging of topsoil than contributions to the American commonwealth. In his 1927 survey of the Pennyroyal province Carl Sauer observed that the Land Between the Rivers had become a "region dominated in landscape by forest, sparsely populated and little visited by outsiders, separated and isolated in its neighborhood life, divided in its outlook, little touched by progress." In the Pennyroyal he noted that "the people have continued to adhere to the soil in unusual measure," even if it required ever smaller and less productive farms, and the reason, he thought, "must be sought in a condition of contentment of the people with their home which is sentimental rather than economic." An intellectual rationalization, even celebration, for that state came in 1930 when the Southern Agrarian movement, headed by John Crowe Ransom, published a paean to blood-and-soil nationalism in a famous manifesto *I'll Take My Stand*. Kneaded with more progressive ideas, especially regarding the South's racial caste system, the notion has persisted, most recently reincarnating in Wendell Berry.[5]

What had once been a portal for unstoppable pioneering over the Appalachians and into what Francis Parkman in his grand history of the imperial contest for North America had called the Great West became notorious as the voice of rural rootedness and for its stubborn resistance to modernity.

=====

What the Mid-American backwoodsman was to national history, American oaks were to the country's ecology. Oak was an informing species that's ranged from the cultural hearth of the backwoods frontier in the upper Delaware to its high-water mark in the Ozark Mountains, an echo of the Cumberland Plateau. Just as hunting combined with other practices to make a frontier economy, so oaks combined with other flora to make the frontier forest. An oak-hickory condominium was particularly productive of wildlife and tolerant of understory grasses, and it was especially amenable to management by burning.

The region's biogeography had plenty of features that left it far from a monoculture. There were many kinds of trees, many species even of oak (50) and hickory (11–12), and many varieties of grassy patches. Hilly, well-watered landscapes abound with nooks and crannies, odd niches and refugia. Ridgetops were subject to harsher winds, ice, and drying sunlight; ravines and bottomlands were more sheltered and moist. Hilly knobs might be grassy, or outright bald, and exposed sheets of rock might hold only a microflora. There were wetlands and lands that seasonally wetted and dried. Broad basins gave winds a long fetch. Limestone cap rock let surface waters sink into subterranean rills and caverns, such that erosion moved below ground, and the surface could parch when rains faltered.

This geographic patchquilt of hills, dales, plains, ravines, and knobs held a lot of variety, though in general they parse into three biomes. The karst plains were grasslands. The ravines and floodplains were dense with wetlands, canebrakes, mesic hardwoods like red maple, yellow poplar, and American beech. The hilly uplands hosted the oak-hickory complex, sometimes dense with woods, often open with grasses and grading into savannas. It's worth noting that this is essentially a geography of burning. The uplands were drained and at least seasonally dry, and could carry fire. The ravines were more inclined to hold moisture and were shielded from the winds. The plains would seasonally flip from surface waters to subsurface streams, and were exposed to a long fetch of winds—the ideal formula for savannas and steppes the world around.

But this formulation describes only the physical matrix. The reality was that people had existed on the scene for thousands of years. They were present even as the ice was still retreating, as the modern climate began to congeal, as the throng of species released by the inflection from Pleistocene to Holocene began the ecological frontier settlement of North America. Those biomes did not organize themselves and then endure the onslaught of humans through a succession of more powerful technologies and brutal occupations. They coevolved with people. They were partly cultural landscapes; they represented the mutual action of natural features and human societies on each other. When the people left, or when one people replaced another, those landscapes morphed into new arrangements.

Not only the essence but the strength of that interaction changed. Compared with the climatic conditions that had overrun so much of North America with ice, human hunters were feeble. Yet the evidence suggests that they were instrumental in the extinction of megafauna, directly through hunting or indirectly through other habitat tinkering, and that the loss of those big beasts could alter biomes independently of Milankovitch cycles. As human capacity grew, so did their imprint on the land. They burned for hunting and foraging and to improve travel. When the climate turned warm during the Hypsithermal, anthropogenic fires were present to quicken and leverage that change over vast landscapes.

As the tendrils of agriculture reached into the southeastern United States and the eastern watersheds of the Mississippi River, manipulation grew. Squash and maize and later beans led to fire clearing in bottomlands for cultivation and to burning in the uplands for hunting and foraging. Weeds increased. Those flora flourished best that could respond to routine disturbances, particularly fire. Gradually, agriculture moved up the slopes. Shifting cultivation became more sedentary. Where fire would free-range, it supported extensive grasslands, which drew game toward them. Then Europeans arrived, and before they themselves settled, their crops, grasses, weeds, hogs, honeybees, and cattle, and of course diseases, were already being stirred into the mix. Throughout, fire had kept the pot boiling.[6]

The detailed reconstruction of such histories comes from the Cumberland Plateau and Little Tennessee River Valley, but there is no reason to believe that the scenario was not general throughout the region, or that the newcomers streaming through the Appalachians did not adapt those fire practices even as they tweaked the biotic matrix within which those fires burned. At a landscape scale the long waves of human demographics were becoming as significant for the shape of landscapes as sunspot cycles and the rhythms of the Atlantic Multidecadal Oscillation.

By the time parties of land-hungry pioneers pushed into the region it was likely already unraveling from its prior state due to an upheaval in demographics. Yet despite overhunting, poor plowing, free-ranging grazing by livestock, plow-and-abandon row cropping, repeated cutting (some parts of the oak woodland are on their fourth forest since American settlement), and a modified regimen of burning, many of the fundamentals remained unchanged. The new grazers like cattle needed grass as much as bison and elk did, and the way to keep it was to burn it. The new browsers like swine fed on mast, typically from oaks, as much as turkeys did. In this way many of the old pieces remained on the site, even if the kaleidoscope of settlement had rearranged them into new patterns.

For decades the new economy had many elements of the old, much as observers noted that the hunting and small-farm economy of the pioneers (and even their clothing) looked like nothing so much as that of the people they had displaced. Hunting and trapping endured, though the preferred game went from bears, bison, and beaver to raccoons and squirrels. Farmers still fallowed and burned, and when the soil became exhausted, they moved to another plot and did it again. Open-range herding remained legal until the late 1940s, despite its abuses. New forbs and grasses like timothy shoved aside big bluestem and Indian grass. The woods were cut and regrown, and cut and regrown again, and again, though not until the advent of steam did landscape-scale logging become truly established. So, too, not until an industrial economy replaced the older one did open burning rapidly recede and a new woodland thick with mesic hardwoods crowd into fallowed land and rough pasture and overwhelm barrens to create the densely forested landscapes of recent decades, scenes that resemble the past only in broad brush and pieces. The biota, it seemed, also had its trail of tears.

What is the region's true state of nature? It's a metaphysical question. It certainly wouldn't be the scene viewed when Christopher Gist and Dr. Thomas Walker pushed over the mountains, or when James Harrod and Daniel Boone platted their stockades, because indigenous burning had kept the land rich with the savannas, barrens, and glades that drew the pioneers on. And it certainly wouldn't be the land today even if it were left utterly alone because too many species have been introduced and the land repeatedly overturned, from its woods to its soils, by 200 years of human finagling. "State of nature" is a question that matters to Rousseauean philosophers and wilderness purists. It's an impractical question to partisans of biodiversity and ecosystem health. The issue is how best to manage what legacy landscapes now exist, spared from the rudest cutting, plowing, hunting, herding, and burning.

So, what is the region's fire history? And what would make an appropriate fire regime for today, and the anticipated Anthropocene future? It would be surprising if fire history veered radically from landscape history. Fire integrated how people lived on the land: it would be awkward to the point of destabilizing if people lived one way and used fire another. Instead, fire remained both pervasive and particular. Few patches were untouched, though not all were fired routinely. Grasses burned frequently, oak-hickory woods somewhat less so, and sheltered moist gullies only now and again.

These were, after all, mixed biomes, and fire is a profoundly interactive event that takes its character from its context. Even amid the clan of oaks, assorted species vary significantly in how and with what consequences they burn; and within a single species, timing, intensity, and frequency can yield different outcomes. The story of the Barrens—a shallow-rooted grassland, really—is broadly clear in that without fire the land soon overgrew with woods and shrubs. The story of oak woodlands is more complicated. Acorns, seedlings, saplings, mature trees—at each stage in its life cycle, a particular species could respond differently to a fire; to the fire regime according to whether fires burned through the dormant or the growing season, whether one fire followed another the next year or the next decade, whether a fire burned by heading or backing; to the history of disturbances by wind, ice, flooding, drought, logging, grazing, plowing; and to the complex of surrounding species, whether natural or exotic, pyrophobic or pyrophilic. The permutations can resemble the ecological equivalent of a Rubik's Cube. The one surety is that fire was an integral feature to those landscapes and that, as the landscapes reincarnated, so fire assumed new avatars.[7]

As native tribes succeeded to Euro-Americans, fire morphed into new regimes, but not so much that it vanished in significant ways. Fire was widespread but ordinary, and rarely savage. By the early 20th century the land was sagging from cumulative use, a kind of ecological repetitive-stress injury. Like the landscape that sustained it, like the earth left after rains carried the best soils down hillside furrows, fire was fatigued. Landscape burning was a relic practice and increasingly a scorned one. Fire's biotic partners were much diminished or gone. It became increasingly invisible, like folk handcrafts, or something lost in the hollows. The oak woodlands settled into a backwater eddy amid America's bustling pyrogeography. Woodsburning might still exist, like moonshining, but the assumption among the elite was that it could go away and no one, and certainly no biome, would miss it.

A few years after the Southern Agrarians celebrated the region's cultural isolation from urban and industrial norms, the Great Depression began to reconnect it to national standards. The New Deal brought in the Works Progress Administration, the Civilian Conservation Corps, the Soil Conservation Service, and the Tennessee Valley Authority to build infrastructure and rehabilitate land. But if those efforts caused flames to seep away from public lands, of which there were few, fire continued to steep the rural countryside. Hunting and herding persisted in the old way through the late 1940s.

In December 1959 a fire conference to promote protection for southern forests met in Louisville, chosen because earlier that year Kentucky had passed legislation that made statewide protection possible, although funding made the prospect quixotic. Some 3.5 million acres in 50 Kentucky counties (out of 120) remained outside the system, which the conferees were told amounted to "one-tenth of all the remaining unprotected forest land in the United States." Still, the conference agreed with U.S. Forest Service chief Richard McArdle that the situation was relatively simple because 99 percent of fires started from people and "people are easier to control than lightning." Only to a professional forester of McArdle's generation could such a comment seem self-evident.[8]

Then inherited wisdom began a slow implosion. Three years after the Louisville gathering Tall Timbers Research Station held its inaugural fire ecology conference and the Nature Conservancy kindled its natal prescribed burn—together the first shots fired in a fire revolution. The oak woodlands had largely missed the country's first wave of fire protection. The question was whether it would also miss the second wave of fire management.

Chief McArdle was right about people being at the core. Fire's management in the region was about people, which meant about culture, which meant fire management pivoted on the administration of anthropogenic landscapes. In that first era of fire protection most advocates, for or against, had argued over suitable institutions. In the second era the fights would circle more over ideas. It made no sense to push for fire's restoration if, to scientists, land managers, and policy wonks, fire didn't really belong. Someone would have to lead American civilization through the obscure gaps in existing assumptions about what lands were worth preserving and about what role fire should play. The fire revolution, it seems, also had its long hunters, and it was their destiny to lead a generation of pioneers to new lands quite unlike anything most people had expected.

A DARK AND BURNING GROUND

WHAT PROPELLED AMERICAN settlement was land. The economics behind the surge was land speculation, writs for land granted Revolutionary War veterans, and simple land hunger from people with big families and little money. But the land was only as good as its ability to support even a primitive agriculture. So they passed through the mountains and settled in the woodland savannas and clustered around the fabled hunting grounds they called Barrens.

"Kentucke" became notorious as a contested land between Shawnee and Cherokee, Britain and France, Britain and its North American colonists, Americans and everyone else. Even when war and treaty had ceded the land to the new American republic, land tenure was a shambles because no controlling authority oversaw the process. States issued grants, speculators sold unseen lots, and squatters, often in good conscience, claimed plots without adequate survey or legal title—often the same lands. Legal fighting could take years and exhaust the poor, who moved on.

But no less contested, among natural historians and intellectuals generally, was the character of those savannas and their origin. Though pioneers seemed to advance with a rifle in one hand and an ax in the other, much of the landscape was grassy—that's what allowed passage and fed game and pastured cattle. West of the Cumberland Plateau the trees thinned and crowded into sheltered refugia and those grassy openings broadened into swathes, the outliers of North America's tallgrass prairie, breaking and thinning atop shallow limestone. Settlers pondered the relative fertility of the soil. Elites argued over their origin. In the history of geographic ideas, the Barrens became a dark and bloody ground.[1] Here was an intellectual equivalent of edge effect. One side viewed the scene from the woods and insisted that landscapes on such a scale and inhabited by "savages" (literally, people of the forest who lived "by the chase") could only come from comparably scaled natural forces, notably climate, as synthesized by the interaction of climate on soil or perhaps through recurrent flooding. Dense forests did not exist because the natural geography, organized by climate, could not support them. The other side viewed the scene from the grasslands and argued that the landscape was the outcome of burning by America's aboriginal peoples. Those lands could, in fact, support trees, and they quickly did as soon as the fires ceased. Because those innumerable and informing fires were set by people, the barrens were a created landscape.

In their tangled confusions, the rival claims could have put Kentucky's early land courts to shame. Like them, too, the dispute tended to divide between those on the ground and those in offices, courts, and academies. But just as the long hunters and collateral pioneers moved on, so did the discourse over origins. It just changed names, handing down the legacy of one generation to its successor. It continues today, and it continues to entangle life on the ground, particularly the practice of managing fire.

———————

The great Barrens built on three geologic features. They were relatively level, they sat atop karstified limestone that drained surface waters away, and they had outcroppings of salt. The latter made them attractive to megafauna, which attracted big-game hunters who followed the trails the game made to the salt licks. The hunters set fires to help keep the landscape open and in the kind of browse and grass the game liked. The karst surface seasonally emptied the land of the dissecting streams and wetlands that elsewhere prevented fire from spreading. The rolling topography allowed winds to carry fire until the grasses gave out amid rock and ravine. People kept fire constant: geography decided how fire could propagate.

The first English speakers to meet them, widening out from the hills and forests, called them "barrens" or sometimes "deserts." The terms reflected more on English etymology than on geography. A "desert" was not necessarily an arid place, but one deserted of people. A "barren" was not intrinsically sterile, but simply empty of woods. (For that matter a "forest" was not a biological term at all but a legal one, derived from Norman law, that defined land reserved from common use and subject to forest law.) Eventually English borrowed the term "prairie" (from the Latin *pratum*, meadow) from French explorers. But the earliest travelers reached for a word that would describe a land empty of permanent woods and sedentary peoples.

Universally, they describe the land as fertile and lush with pasture. The absence of wood was a problem for construction and fuel, but the barrens made a splendid hunting ground and pasture; the problem was to settle lands that allowed access to both. Uniformly, too, observers noted the fires that gathered on the barrens along with the bison. The classic account remains François André Michaux's published in 1805, who described their character: "Every year, in the course of the months of March or April, the inhabitants set fire to the grass. . . . The custom of burning meadows was formerly practiced by the natives, who came in part to this part of the country to hunt; in fact, they do it now in the other parts of North America, where there are *savannahs* of an immense extent." Then their purpose: "Their aim in setting fire to it is to allure the stags, bisons, etc., into the part which are burnt, where they can discern them at a greater distance." And then their potential ferocity: "Unless a person has seen these dreadful conflagrations, it is impossible to form the least idea of them. The flames occupy generally an extent of several miles, are sometimes driven by the wind with such rapidity, that the inhabitants, even on horseback, have become prey to them."[2]

Michaux speculated that a "great conflagration" had birthed the "spacious meadows in Kentucky and Tennessee." Certainly, without fire, they could not persist. "When chance prevents any part from the ravages of the flame, for a certain number of years they are re-stocked with trees," but those "extremely thick" woods make another conflagration likely, which returns them to "a sort of meadow." If ignition varied, as with a climatic source, the woods and barrens would ebb and flow. But if ignition was constant, as it was in the hands of the indigenes, the flames would gradually beat back the woods. And where conditions favored more expansive burning, prairie would dominate, as it increasingly did to the west. In brief, the prairie barrens relied on fire to flourish.[3]

Writing in 1819 R. W. Wells elaborated on the mechanics to the readers of the *American Journal of Science and the Arts*. The prairies were found from the Appalachians westward, sometimes inscribing large arcs and peninsulas, until they became the primary landscape.

> The Indians, it is presumed, (and the writer, from a residence in their country and with them, is well acquainted with their customs) burn woods, not *ordinarily* for the purpose of taking or catching game . . . but for many other advantages attending that practice. If the woods be not burned as usual, the hunter finds it impossible to kill the game, which alarmed at the great deal of noise made in walking through the dry grass and leaves, flee in all directions at his approach. Also, the Indians travel much during the winter, from one village to another, and to and from the quantity of briars, vines, grass, etc. To remedy these and many other inconveniences, even the woods were originally burned so as to cause prairies, and for the same and like reasons they continue to be burned towards the close of the Indian summer.[4]

Like Michaux, Wells thought that successive fires had created the barrens. "Woodland is not commonly changed to prairie by one burning, but by several successive conflagrations; the first will kill the undergrowth, which causing a greater opening, and admitting the sun and air more freely, increases the quantity of grass the ensuing season: the conflagration consequently increases, and is sufficiently powerful to destroy the smaller timber; and on the third year you behold an open prairie." He himself had witnessed in the country between the Mississippi and Missouri, "after unusual dry seasons," woodlands transformed into prairie. But the burning had to continue. If it ceased by "accidental causes" or heavy grazing, a young forest would spring. In fact, one could measure proximity to cities by the height of the surrounding trees. The nearer the town, the greater the woods.[5]

Thomas Jefferson also sided with the burners. In corresponding with John Adams about the "usage of hunting in circles" among American Indian tribes, he noted it "has been practiced by them all." But "their numbers not enabling them, like Genghis Khan's seven hundred thousand, to form themselves into circles of an hundred miles diameter, they make their circle by firing the leaves fallen on the ground, which gradually forcing animals to the center, they there slaughter them with arrows, darts, and other missiles." The practice was called "fire hunting," and was picked up by colonists. Jefferson shrewdly suggested that this practice was "the most probable cause of the origin and extension of the vast prairies in the western country."[6]

Those opinions were handed down like woodcraft. By the time trained naturalists arrived in numbers, the land was settled, its barrens reduced or gone, and its fires tamed into a kind of rural domestication. In his *Geological Survey of Kentucky* David Dale Owen relied on the testimony of the "old inhabitants" of the Pennyroyal, who all declared that "when the country was first settled it was, for the most part, an open prairie district, with hardly a stick of timber sufficient to make a rail, as far as the eye could reach, where now forests exist of trees of medium growth, obstructing entirely the view." They "attribute this change to the wild fires which formerly used to sweep over the whole country, in dry seasons, being now, for the most part, avoided or subdued, if by accident they should break out." Later, Nathaniel Shaler, a Kentucky native with the U.S. Geological Survey, observed that "after the Indians were driven away, about 50 years elapsed before the country was generally settled, and in this period the woods to a considerable extent recovered possession of the areas of open ground." The reason: "the habit which the aborigines had of firing the grasses in the open ground." In his study of the Pennyroyal region Carl Sauer spoke simply of the "popular unanimity in ascribing the barrens to fires."[7]

The land changed. Barrens, fires, and old-timers passed away. What remained were the mature woods that, along with farms, had reclaimed the prairies, a population that tended to look back rather than ahead, and endemic woods fires that to outsiders belonged with the typhoid and pellagra that marked a subsistence economy. By the time John Muir crossed the Cumberland in 1867, he reported black oaks, "many of which were sixty or seventy feet in height, and are said to have grown since the fires were kept off, forty years ago." In a reversal of common wisdom pioneering had not cleared the land but encumbered it. In the Barrens the process had not gone from woods to clearing but from clearing to woods.[8]

By the early 20th century, what Betty Joe Wallace said of the Land Between the Rivers could apply to much of the rural oak woodlands. The population had barely recovered from the Civil War, "the people were plagued with communicable diseases," houses and farms "had not changed considerably since the 1870s," coal competed with the fuel wood and lumber industry, transportation was still primitive, and "farmers continued to cultivate their land in much the same manner as their ancestors," which is to say, a long fallow slash-and-burn that left more and more of the land as rough pasture. Even burning seemed a shadow of its former self.[9]

There was no one to recall from living memory what the Barrens and contact-era oak-hickory forest had been like and how fire had rippled across and through them with the insouciance of their former bears and bison. Instead, professional scientists and bureaus pronounced on the scene, and did so in journals and government publications. If they noted fire, they had little good to say about it. Botanists worried about rare species possibly threatened by flame. Entomologists sneered at woodsburners who claimed fire helped control ticks and chiggers. Wildlife biologists worried about spring fires that might overrun nests. Agronomists fretted over a fire-powered economy that had, since ancient times, been condemned as primitive. And, most of all, foresters thundered scorn and damnation on those who argued from folk experience that fire made life and land better.

In 1905 the U.S. Forest Service acquired the nation's forest reserves and began a system of formal fire protection, which academic forestry and colonial exemplars insisted had to begin with fire control. (Bernard Fernow even insisted that fire protection was not truly a part of forestry but a precondition for it.) After the trauma of the Big Blowup of 1910, the Forest Service doubled down in its determination to control fire, and the 1911 Weeks Act, passed while the smoke from the great fires had barely blown away, gave it the power to establish cooperative fire programs with state forestry bureaus. Kentucky joined in 1912, Tennessee in 1914. The official voice of elites and Progressive Era agencies was that woodsburning belonged with uphill plowing, free-range grazing, and swiddened row cropping as part of the slovenly agronomy that had left the region mired in backwardness and poverty. By now forestry had become the official oracle for all matters concerned with fire.

There were some dissenters. In California a gaggle of critics argued for "light-burning," which they regarded as the "Indian way" of forest management, until forestry condemned the concept as anathema. In the South woodsburning remained so pervasive that the Forest Service even hired a psychologist to investigate why ruralites continued to burn when every rational person knew it was damaging (his conclusion: they burned because their "pappies" had). But if fire was harmful to grasses and pines, it was ruinous in hardwoods. There was no official or scientific dissent from the assertion that fire had no place in oak-hickory woodlands. It existed only because ignorant inhabitants in hollows and back forty farms continued to burn out of superstition and a mindless adherence to tradition. Woodsburning belonged with snake handling. It seemed as endemic as typhoid fever.

Woodsburning, or more properly savanna burning, was beyond the pale of professional forestry. In 1959 when Julian Steyermark summarized the flora of the Ozark Mountains, he included a review of fire. He noted that nearly all travelers attributed the prairie patches and open woodlands to routine burning, and he accepted that even a few contemporary botanists have been "impressed by the tales and beliefs" of those early observers. What Indians did the first settlers did as well, "annually" burning over "large tracts of woodland in the fallacious hope that the grass would be improved." All they proved was that repeated burning retarded forest growth.[10]

Apparently evidence at hand mattered less than what forestry's reigning authorities had to say. Most of Steyermark's inquiry quotes and annotates H. M. Raup's 1938 study of southern New England, in which Raup concluded that if the indigenes had burned as often as reported, there would have been no forest, no herbaceous understory, and no usable fauna. "A similar line of reasoning," Steyermark concluded, "would seem evidence for the Ozarks." So contrary to forestry doctrine was the practice of routine burning that Raup was led to dismiss the written accounts as obviously at odds with what academic science knew. The oak-hickory forest that existed was "the normal vegetational expression of the climatic-edaphic complex." Fires could mar those woods and even prevent their growth, but it was absurd to believe that those forests existed because of the prevailing fire regime. "By the same reasoning," Steyermark affirms, "the same general picture holds true for the Ozarks." Rarely has a professional elite been both so thoroughly wrong and so smug in their arrogance (okay, economists in recent years). The unlettered locals and exploring naturalists had it right.[11]

The cultural lore of fire shifted from folk practitioners to intellectuals. Academic knowledge was to folk knowledge as industrialization was to subsistence farming and herding. Botanists, and especially foresters, denounced fire: it had to go if the region was to recover from its slow wasting disease and modernize. That fire had historically occurred meant little; so had market trapping. That did not make raccoon pelts the basis for a modern economy. Old guard state forestry bureaus, which were founded on fire protection, remained a bastion of pyrophobia.

Yet there was one notable dissenter. Carl Sauer had grown up in German communities in Warrenton, Missouri. When he was 10, he was sent to Calw, Germany, for two years to learn the proper discipline to become a scholar. Graduating from Central Wesleyan College in 1908, he went to the University of Chicago and left as a geographer; his European training meant that, for him, geography was also history because landscapes were cultural creations. Later, he spent long field seasons in Mexico, living as an ethnographer more than a classic scholar. By the time he retired from the University of California, he had established America's greatest school of historical geography.[12]

Sauer began his career in his home region. His dissertation focused on the Ozarks; he wrote surveys of the Illinois prairie peninsula and the Pennyroyal barrens of Kentucky. In 1919 he established a field school near Mill Springs in south-central Kentucky, where the Cumberland Plateau smooths and grades into the karstified plains and the old barrens—a microcosm of the region. He knew fire as a living practice—had seen it around him as he grew up, had found it in the historical records of the regions he studied, had undoubtedly visited Mexico under palls of seasonal smoke. Unlike others, particularly foresters, he did not see those flames through the prism of authoritative texts but with his own eyes.

He sensed the pervasiveness of fire and its power, particularly for pre-industrial cultures. He didn't pull punches in describing the stilted state of Ozark and Pennyroyal society, but neither did he dismiss all their practices as meaningless or damaging; he was an ethnographer who synthesized his observations into geography and who sought to understand how people and land had interacted to shape a scene. He once observed that he felt more at home in the world of the past, an agricultural world built of living materials, than he did in the industrial world he had grown into as an adult (in fact, agricultural origins became one of his major contributions), and he preferred local cultures to powerful capitals and colonial aspirations. That different view of human life gave him a different view of fire. Unlike virtually all his contemporaries he didn't condemn fire. He sought to understand how it worked.

In his Pennyroyal survey published in 1927 he observed that "we have here probably an illustration on a large scale of the fact that even primitive man is able to modify profoundly his environment by the aid of his more powerful tool, fire." He suggested a parallel with the Ozarks, "which also show the same feature of deforested upland flats." He speculated that such dynamics were probably common to all grasslands, and that anthropogenic grasslands testified to the fact "that wherever primitive man has had the opportunity to turn fire loose on the land, he seems to have done so," and that "having fire, man was enabled to go forward to possess the world."[13]

His thinking about fire climaxed in the celebrated 1954 symposium, "Man's Role in Changing the Face of the Earth." One of the three organizers, he offered a judgment based on a lifetime of scholarship. Speech, tools, and fire, he reckoned, were the "tripod of culture and have been so, we think, from the beginning." As a student, James Parsons, summarized in Sauer's obituary, "He early and persistently pointed to the antiquity of fire as a human tool" and "the modification of vegetation cover by burning he saw as a continuous process covering the entire history of human presence on the earth." He believed that "in areas controlled by customary burning, a near-ecologic equilibrium may have been attained, a biotic recombination maintained by similarly repeated human intervention." Besides, he noted, complete protection from fire was self-destructive, and suggested that "the question is now being raised whether well-regulated fires may not have an ecologic role beneficent to modern man, as they did in older days." By "the possession of fire," a transmutational power, humanity "was able to enter new environments and was securely started on his way towards dominating and modifying the world, the world of nature to become that of culture."[14]

By the 1930s querulous critics began to pester the orthodoxy that fire was an unmitigated environmental evil. S. W. Greene, Herbert Stoddard, E. V. Komarek, Omer Stewart, later prairie enthusiasts like John Curtis, even an occasional renegade forester like H. H. Chapman challenged prevailing authorities to argue for fire's utility. But all were from the southeastern United States, save for Stewart, who studied the western Paiutes, and Curtis, who wanted to restore tallgrass prairie in Wisconsin. In the early 1940s foresters had reconciled themselves to a degree of southern exceptionalism for fire in pines, notably the longleaf, but they held their ground regarding hardwoods. It had no legitimate place in the dense hardwoods of the Northeast, nor in the oak-hickory complex. By the time Sauer was concluding that humanity owed much of its earth-shaping power to fire, professional forestry had quarantined it into subregional niches and subcultures. By the time *Man's Role in Changing the Face of the Earth* was published in 1956, America was in a cold war on fire, and three years later the Louisville fire conference identified rural Kentucky, in particular, as desperately deficient in forest fire protection.

The fire revolution passed the oak woodlands by. Researchers argued for a place for fire in the pitch pine on southern New Jersey barrens and in prairie patches of the upper Midwest and the old pineries from the montane West to southeastern coastal plain. But no one spoke for the diminished but regrown oak-hickory-barrens landscapes as an arena for fire. There were no partisans to speak for oak as there were for longleaf, ponderosa, lodgepole, and sequoias. There was no reason to because everyone knew that fire had no place in hardwoods, that the only fires that occurred were incidental to the regional ecology, and Carl Sauer was not a member of the disciplinary and bureaucratic tribes fighting over the role, beneficent or malevolent, for fire in the region.

In 1957 he retired. He had long since moved on to questions regarding Mexico and early humans, but the imprinting of those years in the oak woodlands never faded, so let him have a final word. It was by humanity's "distribution through all climatic regions and his power to employ great physical forces"—notably fire—that "man has become the guide fossil of the present geologic period." If the field station at Mills Springs outside Monticello, Kentucky, was a microcosm of the region, the oak woodlands and their peripheral prairies had become, for Sauer, a microcosm of the Earth.[15]

The fire revolution stalled during the 1980s, then renewed in the mid-1990s. During that same becalmed period, it became apparent to serious observers that the traditional hardwoods of the region, most ostensibly its oaks, were fading. They weren't reproducing. They weren't thriving relative to hardy invasives like eastern red cedar, aggressive exotics like autumn olive, and more mesic species such as red maple and American beech that had formerly been confined to ravines and wetlands. They were no longer functioning as keystone species for wildlife. Then, about the same time as the fire revolution revived, Marc Abrams proposed that the reason behind the shift was the loss of fire. Here was the old debate about the barrens reincarnated.[16]

The oak-fire hypothesis, as it became known, argued that fire had not only been historically prevalent in the oak woodlands and affiliated savannas but that it had been essential to the ecological integrity of the biome and that fire's rapid removal in recent decades was a disturbance that threatened to drive many oaks to the brink of exhaustion, if not extinction. When first announced it was a startling proposition. The region lay outside the main provinces of American fire. Its fire institutions were pretty sparse on the ground. No one seriously studied fire because, it was understood, there was no fire ecology to study. There was nothing to be learned. Hardwoods had no place for fire. The region's annoying, residual fire habits would gradually fade away like candles and kerosene lamps in the presence of electricity. If you drew a map of American fire, the oak woodlands would be the empty vacuum between the anode of southern burning and the cathode of historic North Woods conflagrations. Intellectual understanding mirrored that distribution. No one published because there was nothing academically interesting to publish about. Everyone knew fire had no useful place in hardwoods; and since there was no ecologically useful fire history to be found, there was no point in looking.

The oak-fire hypothesis said otherwise. The old debate about fire and prairie was revived in its fundamentals. Critics quickly declared that correlation did not mean causation. That fires had occurred did not mean they were informing; that the biota had survived fire did not mean fire was essential to its survival. But as happened before, the pyromantics won. When dendrochronologists studied fire-scarred trees, they found a long and robust history of burning. When silviculturists experimented with acorn regeneration, they found fire useful. If it was not possible to fully restore the past, it might be possible to nudge the present scene toward a brighter future. With a renewed fire revolution to fill its sails, prescribed fire became the treatment of choice for restoring health to an infirm biome.

Twenty years later a richer sense of eco-complexity has qualified the early enthusiasms. No one outside of old-school foresters rejected fire's presence on the landscape: they just wanted more nuance. Fire is not a precision instrument but a broad-spectrum ecological catalyst, and there was more to the hardwood forest than its cohort score of oaks. More nuanced prescriptions were needed that could capture the diffuse interactions that fire set into motion. Not everyone wanted oaks primarily; what was good for oak, and what it could tolerate, might be poor for particular species of invertebrates or birds. Even for the oaks it mattered when, and with what intensity, fires burned relative to critical moments in a tree's life history, from pollination to germination, seedling development, and release into the canopy. Moreover, the sites being treated were often so tiny that they couldn't tolerate broadcast burning; managers couldn't substitute breadth of land for variety of effects. The oak-fire hypothesis endures. What is debated is how, exactly, to best apply its insights.[17]

At the Great Council at Sycamore Shoals in 1775, Cherokee chief Dragging Canoe is generally credited with introducing the phrase that there was a "dark cloud" over Kentucky. It has been widely interpreted that he referred to ancient quarrels over ownership to its opulent hunting grounds. But there might have been a special piquancy to the allusion because, in March, when his oration was given, the skies would have been dark with the smoke from traditional rites of burning.

UNCHANGED PAST

Stones River National Battlefield

O N DECEMBER 30, 1862, the Army of the Cumberland under General William Rosecrans faced the Army of Tennessee under General Braxton Bragg at Stones River, outside Murfreesboro, Tennessee. The next morning the Confederates launched a dawn attack that surprised and so routed the Union army's right flank that the Army of the Cumberland threatened to roll up on itself. Some ground scabby with exposed limestone and some cedar forests, along with the scrambled sprawl of the assault, helped stall the surge until Union troops rallied under General Philip Sheridan, found high ground, and eventually returned a savage artillery fire that blunted and then broke the Confederate attack. The next day, with a major victory in hand, President Abraham Lincoln issued the Emancipation Proclamation.[1]

In 1927 Congress designated the site as a national battlefield. Today the National Park Service (NPS) oversees it. The actual landscape mattered on the day of battle, and it matters to those charged with curating the battlefield as a historic site. Their charge is an unusual one for a land agency: to maintain the field as closely as possible to what it was on December 31, 1862.

But Stones River is not a work in stone, like the Hazen Brigade Monument that commemorates the only Union brigade not dislodged at some point by Confederate troops (it's the oldest intact Civil War monument in the nation). Stones River is a living landscape atop a scene blotchy with craggy limestone exposures. Instead of worrying about the weathering of marble, the NPS must fret over a biotic landscape that threatens to overrun the scene as Confederate divisions did the Federal's right flank. Trees and grasses grow, habitats entangle with invasives and weeds and rearrange themselves, cedar groves thicken and become more impenetrable than when they disrupted the Confederate charge, and what were open fields of fire quickly close with all the vegetation that makes the site a patchwork of rare species and green sludge.

Besides that mandate, the land itself—not farmed or paved into suburbs—has collateral properties that make it a place for protecting rare and threatened species and that serves as an urban park for recreation by residents of Murfreesboro. To its credit the park staff has accepted those ancillary duties and has sought methods that can both clear and rejuvenate. The battle was saved for the Union army by the firepower of its artillery. The battlefield might be saved for posterity by a more benign ecological firepower.

———

In 1862 the landscape was fundamentally a karstified limestone, patchy with exposed rock and cedar glades. The stone and trees were enough to break up a unified charge, which probably spared the Union army, but most of the scene was relatively open either as prairie or woodland savanna. Much of that scene remains, but only in its gross arrangement. The cedars have burst out of their nooks to join mesic woods, the grasslands overgrow with woody weeds if those invasives are not forcibly removed, and the setting seems as much an urban park as a memorial.

The major fields were kept open by mowing. In the 1980s park staff experimented with fire. Something resembling a program began after the National Fire Plan arrived. The burning brought other beneficial side effects, particularly ecological, to a region whose protected landscapes tend to be the size of postage stamps. But it also had some less desirable ones, such as smoke and the need for a burning program, which is more bureaucratically complicated than simple mowing. A prescribed fire program requires staffing, planning, and funding, and it has to be done routinely enough to produce the effects wanted. It takes determination to push back against the mounting pressures to let it all go.

In all this, Stones River is a cameo of the region's fire issues. The sites are small, and must be parsed into still smaller pieces for actual burning; it makes for a lot of edge relative to area, which reduces economies of scale. To drive back the encroaching mesics, fires must be hot and often, but proximity to town and structures (there is even a hospital within the airshed) argues for quieter, less frequent burns; the small park staff, geared toward historians, means outside support is needed, which increases the transaction costs; and while some species flourish after fire—the once-threatened Tennessee coneflower so thrives on the burned battlefield that it has been delisted—others may suffer, or are asserted to suffer. The threatened Indiana bat shuts off the calendar for burning on March 31. Red cedar has so rooted that fire must interact with other means, such as cutting or herbicides, in order to cull it. The surrounding city and untended lands are a continual source of infection for invasives and weeds, and some trees, like dogwood, while exotic locally, are liked by urban visitors and so must be kept for aesthetic reasons. Wind storms, ice storms, droughts—all overturn burning schedules. There are no charismatic species to push for fire.

What Stones River does have is an unambiguous mission, to preserve the scene, and a major organization, the National Park Service, to help achieve it. It's part of a circuit of Park Service burning operating around Natchez Trace. Its quest may be quixotic, but its charge provides some administrative and legal leverage not available to most of the other fire-hungry sites in the regional archipelago. Some miles to the south William Faulkner famously declared that the past was never over, it was not even past. But that's not a dictum nature understands. It moves on. With prescribed fire Stones River is trying to let it move while holding those parts of the past that matter most. The past may be with us always, still informing the present, but it is not a sure guide to the future.

The Union victory allowed the Army of the Cumberland to build Fortress Rosecrans at Murfreesboro, which became a major supply and transportation center and contributed to Union control of the west. Stones River can't serve a comparable role for restoring fire through the oak woodlands, but it can serve as a reminder of what restoration can mean and cost, and as a monument to what a determined group can do if it holds its ground.

UNCERTAIN FUTURE

Land Between the Lakes

THE HEARTWOOD OF the oak-hickory region is bounded east and west by the Appalachian Mountains and the Great Plains, and north and south by major rivers, notably the Ohio and the Tennessee, between which flows the Cumberland. By a quirk of glacial geography, the Tennessee River flexes sharply north where Mississippi, Alabama, and Tennessee meet, and then runs parallel to the Mississippi; the Cumberland does the same; and the two rivers flow nearly side by side, with 8–10 miles between them, for 40 miles before they merge and spill into the Ohio. Later, the rivers became lakes after the Tennesse Valley Authority erected dams. That hydrographic oddity created an isthmus, effectively an island, a microcosm of the region. In its institutional history the Land Between the Lakes (LBL) can serve as a cameo for the oak woodlands overall.[1]

Where Stones River National Battlefield showed the perplexities of using the past as a guide, the Land Between the Lakes speaks to the equally vexing problem of using the future. The future has proved mercurial, as unstable as history, elusive in purpose, and accordingly uncertain in how to fix that vision in institutions.

———

It was originally known as the Land Between the Rivers. The rivers tended to isolate rather than connect. Iron deposits and abundant timber (fuel wood) led to a small boom, but mostly progress went west and down the rivers. The Civil War and its aftermath left the place wounded and its economy trashed. The Land Between the Rivers became as isolated within the region as the region was within the country. In 1927 Carl Sauer, who knew it as the Interfluvial Area, called it a place "dominated in landscape by forest, sparsely populated and little visited by outsiders, separated and isolated in its neighborhood life, divided in its outlook, little touched by progress."[2]

The New Deal broke that lethargy with a raft of relief programs, and with the advent of the Tennessee Valley Authority (TVA), it promised what the region had never known, a reasonably comprehensive infrastructure that could connect its parts. These not only made up for the past with roads but pointed to a future in the form of electrical power. A grand experiment in regional planning, the TVA would serve as an economic flywheel; and it would persist beyond the New Deal. Initially its power came from dams, the dams required stable watersheds, and watersheds demanded soil conservation, reforestation, and fire protection. In theory fire institutions might follow the dendritic order of its streams and spread over the countryside.

In 1937 the Department of the Interior acquired 65,000 acres for the Kentucky Woodlands National Wildlife Refuge in the Land Between the Rivers. In 1944 the TVA completed the Kentucky Dam, flooding the western edge and leaving a land between a lake and a river. In 1959 a complementary dam on the Cumberland River was contemplated, which prompted planning for the use of the resulting isthmus between lakes. In 1964 the TVA assumed control over the wildlife refuge and began acquiring further land to fashion a model of outdoor recreation and conservation education, the Land Between the Lakes Recreation Area. Logging and hunting would continue; Austin Peay State University conducted field surveys and created what became a camp for field biology. A year later Lake Barkley began to fill, and the Corps of Engineers cut a canal to join the lakes where they were narrowest. The land had never been so isolated—nor so full of promise to bond with larger trends.

Yet its designated identity as a recreation site faltered—few people came, and as numbers proved disappointing the TVA hesitated to invest more, which set up a downward spiral. What did flourish was its interest to the environmental movement. The Society of American Foresters identified four research natural areas for protection. A network of ecology study areas—34 sites on 500 acres—was created. The National Science Foundation designated LBL and Murray State University's Hancock Biological Station as an experimental ecological reserve. An elk restoration area (which subsequently included bison) was established, later followed by experimental ponds for research. In 1991 LBL become the 47th biosphere reserve in the United States, and the 300th in the world. This required the legal designation of "core areas," which effectively removed them (25 percent of the park in all) from active recreation. By now the TVA regarded LBL as a distraction from its primary mission, electrical power, which was itself stumbling from hydro dams to nuclear plants. There was little enthusiasm for the kind of futuristic planning that had first inspired the project; and there was less money. The public never responded to that founding vision or its piecemeal implementation. Instead of a model of government rehabilitation, LBL became, for many displaced locals and critics, a model of large-scale planning run into a swamp.[3]

A geographer from the University of Tennessee, Ronald Foresta, wrote a study of LBL that carried the subtitle "a geography of the forgotten future." The TVA had invited him to review the site in 1984. Foresta found an appalling "emptiness" to the land, a "vast place" that "seemed utterly lacking in wider context or larger purpose." He observed icily that "two decades" after it had agreed to create the Land Between the Lakes Recreation Area, the "TVA had no institutional memory of why it had created the place." Nor did anyone else. "The degree of forgetting was startling. It was as if the place were cut off from the rest of he world by two branches of the Lethe rather than two tributaries of the Ohio." He concluded that, in its pith, the problem was that LBL had been designed for a future that never arrived. Instead, what remained were the deposits of failed experiments. Land Between the Lakes was to organizing institutions what Big Bone Lick was to Pleistocene megafauna.[4]

In 1999, after the TVA failed to garner the appropriation necessary to manage its holding, LBL was transferred to the U.S. Forest Service. The Forest Service undertook yet another inventory, in the form of an environmental assessment, as it sought to craft working plans. LBL resembled an orphaned nephew, handed around an extended clan. A skeptic might note that what is said of fusion power, that it is a technology of the future and always will be, might be said of LBL's land management mission.

Constraints are many. Of 170,000 acres, some 40,000 are isolated as core areas for the biosphere reserve. Large swaths were committed to the residual intention as recreation, to which were later added research and educational facilities. There is a living history museum. There are elk and bison ranges. There are historic sites, from the stones that once delimited cabins to charcoal furnaces for the briefly flourishing iron industry of antebellum times. Wind and ice storms slash woods, changing the dynamic of fire. The Forest Service has proceeded cautiously with its unsought bequest.

Whatever is done will likely involve fire. Unlike most of the agency's western forests, LBL has no natural fire, and unlike its southeastern holdings there is no continuity of woodsburning adapted into prescribed fire. Yet what visitors most want is wildlife, whether along scenic drives or by hunting. Revealingly, this was the primary use before federal intervention, and it was, in the form of the Kentucky Woodlands National Wildlife Refuge (NWR), the purpose of the first federal reserve. But the birds, bison, elk, and deer flourish best in open sites rich in grass and browse. Historically, this meant hardwood savannas and prairie patches, and fire was the principal means to sustain them. Today, it still means fire, although fires must interact with other treatments and within a social and legal context far removed from historic times. Outside a few sites fire seems unable, unaided, to restore open woodlands fluffed with grasses; and instead of arguing for routine fire, privileged species such as the endangered Indiana bat are squeezing the almanac of burning. In all this there is little unique about the challenges the fire program at LBL faces.

The burning is targeted. Fire officers patch-burn the elk-bison prairies. They burn in two demonstration areas, north and south, about 9,000 acres in all, for oak savanna renewal. They have tried prescribed fire alone as a control and fire with thinning, some of it aggressive. Since 2008 the program burns about 5,000–6,000 acres a year. They would like to do more; they reckon they need at least 20,000–25,000 acres but lack capacity. Besides, the heavy thinning grades into logging, which has aroused local opposition—lingering resentment over TVA's creation of LBL makes anything done by whatever agency oversees the land open to protest. By regional standards the burning is significant and relatively concentrated. By generally accepted ecological standards it falls well below what is needed. Like the rest of LBL, the fire program is something people visit, not a rallying point for the future.

There are many possible interpretations of what went wrong—every reviewer has a favorite—and plenty of musings about what the unstable history of the Land Between the Lakes means. But that institutional fickleness, in fact, may be its primary message to fire history and management. None of its fire issues are unique: they are characteristic of the wider oak woodland region. What they show, however, is the awkwardness of trying to impose a larger institutional order on the land that allows for landscape-scale operations. Even LBL cannot do all it wishes on its own estate because of limited capacity and fragmentation in purpose.

There is no regional matrix by which to hold the scattered parts of fire programs to a common cause and collective action. The TVA filled an institutional vacuum, and then left one. The small holdings, the folk conservatism, and the fierce if peculiar character of independence, which for the Southern Agrarians were a source of pride, can frustrate programs that need to coordinate among and span landscapes and move into the future. The repeated disappointments of LBL are a miniature of the regional search for an organizing mechanism. Without such a matrix the preserved sites, and their fires, may become a cabinet of curiosities in a living history museum rather than serve as points of positive infection by which to renew a living landscape.

UNSETTLED PRESENT

Nature Conservation

THEY ARE MOSTLY SMALL, mostly invisible to the public, and when viewed, they are not seen as fire places. They are landscape patches like Crab Orchard and Clarks' River National Wildlife Refuges; state natural heritage areas like the Jim Scutter, the Raymond Athey, and the Eastview Barrens State Nature Preserves; the Trail of Tears demonstration forest; Nature Conservancy preserves like Mantle Rock; and restoration projects on the Shawnee National Forest. Within the oak woodlands federal lands, and so a federal presence, are sparse: 1.1 percent of Illinois is public, 4.2 percent of Kentucky (mostly in Department of Defense holdings and the Appalachians), 4.8 percent of Tennessee, and except for military installations and the odd park like Mammoth Cave, they are a fine-grained patchquilt of public and private holdings. Isolated, the protected lands appear as archipelagos; a few stand out, the equivalent of high islands like Tahiti, but most more resemble low isles, or even the empty rings of atoll reefs. They are easy to overlook. The public is generally interested in them only when something goes wrong.

But if tiny they are far from trivial. They are the sites where the fire ecology of the oak woodlands today is being discovered, where appropriate fire practices are being devised and applied, where fire-catalyzed patches are being renewed, if not restored. Those patches can punch above their weight class because they have value as symbols, as final sanctuaries, as rallying points for spreading good fire, and as fulcrums to leverage neighboring lands. Henry Thoreau saw the cosmos in Walden Pond. William Faulkner imagined a moral universe in the "little postage-stamp" of a place he called Yoknapatawpha County.

They share a biome, a common history of human land use, and a sense of urgency lest those patches shrivel and blow away. They need what all such operations do. They need legal space, they need staff, they need equipment, they need calendars that won't clash with climate or conflicting ecological purposes, that won't shut down burning by arbitrary dates. They need an economy of scale. They need a powerful sense of purpose. They are flakes from a common core, split off by the blows of history. Each is unique, yet each is synecdoche, highlighting one feature that can stand for the others.

Mantle Rock Preserve is a 367-acre holding under management by the Nature Conservancy (TNC) that lies within the western fringe of the Pennyroyal, and just outside the historic complex of prairie and woodlands known as the Big Barrens.[1]

Its natural and human history has been continuous; records from pollen, fire scars, archaeology, and written texts exist back to 3,900 BP. Across the Ohio River are Clovis sites. Nearby excavations trace Middle and Late Archaic peoples, and then Early Woodland cultures. French explorers appeared in the mid-18th century; an unsettled era of indigenous migrations and wars followed; and by the 1780s American settlers arrived. In the winter of 1838–39 the Trail of Tears sent a forlorn party of Cherokees through the area, camping at Mantle Rock itself.

A fire history accompanies this chronicle. Written accounts suggest near-annual burning of the glades, uplands, and woodland savannas. By the time American-style agriculture firmed up, the pattern of burning shifted because roads, plowed fields, and pastoralism broke the ability of fires to spread. Burning had to accommodate fallowing; each fragment of the land had to be fired separately. Yet the scarring of red cedars on glades suggests that, beginning around 1800, fire inscribed an unbroken narrative on the land. The frequency was highest from 1850 to 1899, the classic "wave of fire" that accompanied landclearing and the replacement of indigenous land uses with those characteristic of the American frontier. Settlers burned to clean up the land—control ticks, chiggers, snakes; promote green up in pasture and rough browse in the woods; stimulate the kind of habitat in which game birds, fur-bearing mammals, and bigger game like deer could flourish; bring arable land in and out of fallow.[2]

But if burning morphed, it persisted. The glades had sparser fuels, and to be recorded the flames had to be timed to the rhythms of red cedar establishment, leaving scars one to two times every 20 years. Outside the glades proper, a richer covering of grasses and woodlands recommend a higher abundance of fire. The most remarkable feature of the record, however, may be its consistency across 150 years. There is little reason to believe early pioneers did not adopt the fire habits of the native peoples here as they did elsewhere; they only tweaked its purposes and leveraged the flames by adding landclearing, part of a crest of fire that tracks the westward movement. Fire changed. Fire endured.[3]

Today's managers don't doubt that fire was and remains a driver of the system. The upland sites are treated as tallgrass prairie, burned every other year. The oak woodlands are burned as possible—not enough to push the biome into true oak savanna, but sufficient to keep it from being overrun by mesic trees and invasives. Cedar glades burn infrequently. The Nature Conservancy uses Mantle Rock to encourage neighboring landowners to restore more native grasses and species (such as game birds, like quail)—a working demonstration that can show how fire can resuscitate fatigued land and biomes burdened by mesofication.

Complications are many. Because some land falls under the crop reduction program, they cannot burn more often than every other year. Invasive grasses and trees like autumn olive hover on the margins, ready to seize any disturbed site, such that the openings that promote native grasses also encourage exotics. A 2009 ice storm left slash piles through the woods that fire officers hesitate to burn out of fear that they might sterilize the soil. Public partners have their own calendars, tied to the political climate of funding cycles. There are no threatened or endangered species that compel managers to burn, only species like the Indiana bat that might (or might not) be inconvenienced by burning, despite the general tonic fire gives to the habitat overall.

So there is less fire than needed, but more than would exist without the commitment and active measures taken.

When Boone and company crossed into Kentucke, the Big Barrens were a defining feature. Today, they exist only in dispersed fragments, the detritus of former times, the biotic equivalent of scattered lithics. By 2014 the Kentucky State Nature Preserves Commission had designated 63 sites for strict protection, of which 14 are barrens in some form or other—glades, barrens, prairie. But it's not enough to set them aside. Unlike lithics they live, they grow, they must be managed if they are to retain the characteristics that led them to be reserved.[4]

Eastview Barrens State Nature Preserve holds 119 oddly configured acres of grassland and open oak woodland. It's co-owned and managed by Kentucky and the Nature Conservancy. Its biodiversity is exceptionally high; it holds grasses and forbs typical of the Barrens and once as pervasive as passenger pigeons, though now driven into enclaves. Among them is a globally imperiled insect.

A nearby railway prone to throwing sparks kept the site more or less burned. But the insect has strict requirements. It can't migrate if the brush (mostly sumac) is higher than four feet, and it needs both burned and unburned patches to thrive. The resolution is not to burn a patch more than every three years, and to demand the same regimen on surrounding patches. In practice this means that any parcel gets burned every five to six years, if everything is aligned, which it almost never is. There is always a drought, a blowdown, an ice storm, a breakdown in collaboration, or a funding crunch to stall a burn. What begins as a planned rotation ends as an opportunistic scramble. The sumac overgrows, the insect is penned to smaller patches, the scale of operations shrinks from 119 acres to a handful. Worse, a good burn opens up a site—that's what it is designed to do, which invites both native and exotic species to move in. Without a second burn, and continuing burning, the site may end in worse shape than before.

Fire management is artisanal—a fire gardening, even a kind of fire horticulture. As co-owner TNC adds some leverage not available to the state; it can, for example, apply for grants to rent a brush mower to clip the sumac (think tractor for farming), but that only adds another set of gears in an already complex mechanism. On a site this small, actually smaller for operational purposes, there is scant room for maneuvering. No surplus space. No discretionary time. Everything seems to work toward shrinking a relic landscape ever smaller. What a bonsai garden is to a working farm, these patches are to a functioning landscape.

What will survive best are those fragments that exist for edaphic reasons. They are true glades, tiny patches of soil on exposed rock. Their preserves are gerrymandered slivers, slabs, and hillsides. Their resident species can take fire, but will survive without it. They are on their own: nature won't kindle fires, and humans, despite their intentions, have failed to set the right kind of fires on a predictable schedule. Legal protection means people won't do bad things, but the social nature of protection means they may find it hard to do good things.

If private landowners and state commissions struggle to operate on the requisite scale, then surely federal agencies can. They tap into national resources and have a support network many times the size required. But as with so much of this biome, the principle is sound, the practice suspect.

The largest landholders are the Department of Defense, the National Park Service, and the U.S. Forest Service. Fort Knox does not consider its surface biota as bullion; Fort Campbell has some spectacular sites, but the reason is inadvertent (an artillery range that constantly starts fires). Mammoth Cave's resources are underground. For a while an interested staff commenced regular burning on its oak woodlands for collateral benefits, much as Stones River National Battlefield does. But the National Park Service is a feudal organization, with each park established by a separate act of Congress and ruled by a baron. When people change, so can purposes. For a few years of the new millennium, the park burned. Then it stopped.

That leaves the Shawnee National Forest. On paper it has the scale so often missing in the region, particularly when aligned with the Cypress Creek and Crab Orchard National Wildlife Refuges and Illinois state parks, forests, wildlife refuges, natural areas, and conservation areas. But up close and personal, it is a micro-mosaic of private and public holdings, more purchase unit than functioning forest, and even the federal land is subdivided to different ends. The Shawnee even has two wilderness areas, the Bald Knob and Clear Springs; the Crab Orchard NWR has one (along with a large Department of Defense [DOD] munitions depot even more shielded). In reality, its geography is another expression of the region's efforts to overcome fragmentation, and its history a chronicle in shorthand of public investments. Yet its size and access to a national institution anchors a complex of public lands.[5]

They sprawl over the Shawnee Hills, a fragment of the Ozark Mountains. Both hills were settled by the same folk in much the same way. But the Shawnee's surface was loess, and clearing, poor plowing, and indifferent husbandry soon slicked the soils off the hills, leaving outcrops like megaliths. So deteriorated was the scene that Illinois passed an enabling act in 1931 that invited the federal government to repossess the land "for forestry purposes." Two years later the New Deal found the federal government reacquiring lands it had previously privatized. The Resettlement Administration moved many farmers off the most abused sites. The Civilian Conservation Corps (CCC) moved in with camps, soil stabilization and reforestation programs, and a form of fire protection. State and federal forests were patched together out of the emptied lands to help create projects for the CCC. In 1939 enough land had been purchased to constitute the Shawnee National Forest. With the cessation of abuses, the land reforested. Not all of it was with native trees; the CCC planted loblolly pine, and invasives infested the scene like spores. But the woodlands that now exist date from the time they were returned to the public domain.

The Shawnee has, officially, a multiple-use mission, but like LBL, its primary purposes are recreation and nature protection. That aligns nicely with the Illinois state sites and the Fish and Wildlife Service. And however pocked and eccentric its holdings, however kinky its borders, the collectivity of public lands creates a condominium with enough mass to undertake projects on more than token sites. That presupposes arrangements to make common needs into collective projects. The feds have interagency agreements; the Forest Service has authorization to cooperate with state forestry bureaus; and the Southern Illinois Prescribed Burn Association joins public and private interests. Southern Illinois University at nearby Carbondale adds academic heft to the cause. The Shawnee can partner with LBL for training. All this gives operational unity to cartographically fragmented lands.

The concern here is mostly what it is everywhere in the region: the oak woodlands are not regenerating and the biota is losing its grassy sheen. Maple, beech, and yellow poplar that formerly crowded into wet ravines and shady nooks have spread widely and grown up beneath the oak canopy. Not only is biodiversity of the sort attractive to ecologists threatened, so is the landscape diversity relished by public visitors who are interested in vistas, game birds, and wildlife, all of which depend on open oakscapes of the kind historically maintained by burning. And it isn't just the uplands: fire histories from Mermet Lake bottomlands show historic frequencies of 1.73 years, with spans from 1 to 15 years, for the years 1895–1965, the era of final reclamation and abandonment. Those burns were fulcrumed by other practices, but the point is clear, the landscape could not be sustained without fire. (By contrast, the Shawnee exhausted five years just to get approval for a single fire, which is to say, it spent two burning cycles just in administrative protocols to prepare to burn.)[6]

There are several experimental sites throughout the patchwork of public lands, but a particularly careful one is the demonstration underway at Trail of Tears State Park. The trial includes four adjacent watersheds totaling 932 acres. One begins with a cut followed by a burn; it has a control site. Another begins with a burn; it, too, has its control plot. The project is part demo—to show the public what opening up the canopy and letting sunlight in can mean in terms of grasses, forbs, butterflies, birds, and oak seedlings. And in part it is an experiment to decide what techniques might best achieve those goals. The project went live in 2013. What it reveals in details may matter less than the simple fact that it is returning fire.

So what do you do?

What do you do if you have an oak forest that flourishes best with frequent fire but is so out of whack that simply dumping fire into it, particularly if you cannot follow up with regular burns, may worsen the scene? If you have a nesting bird (like the woodcock), which means burning is banned after April 15? or a small mammal (like the Indiana bat) that emerges in the spring from wintering caves and whose young can't tolerate smoke? or a flower (take your pick) that can't survive without being burned or by being burned too often? or rare warm-season grasses that respond best to burning in summer? and of course a nearby town that detests smoke?

Worse, the core complication may not be the species that are on site, whether native or exotic, but those that are missing. Millions of passenger pigeons used to flock over the oak woodlands, feeding on acorns. They are gone—and with them whatever biological work they did. Part of why fire does not by itself restore such scenes may be that it is a catalyst; it interacts with other processes. Fire's ecological formulas are not just flame and fuel. There are biological agents at work, and if they are missing, so is part of fire's ecological production.

Historically, the solution was to burn patchily over wide areas, almost year by year. There would always be some places to suit every species. In essence, it's management by abundance. But what if that abundance shrivels to a pittance, such that the patches at risk are small and scattered? In the language of landscape ecology the biome is fragmented, with little chance to geographically join sites on the ground. What bonding exists is institutional. Mostly it means nongovernmental associations to share ideas, personnel, and equipment.

There is another way to imagine the scene. The year before Carl Sauer cohosted the monumental (and never surpassed) symposium "Man's Role in Changing the Face of the Earth," Pulitzer Prize–winning historian David Potter delivered the lectures published as *People of Plenty*, in which he interpreted the American experience through the prism of plentitude. The American system assumed abundance or at least the belief in abundance. There was no need for rationing; there was enough that everyone could find his or her preferred place in the social order and meet his or her own needs. There was ample room on the bus for everyone to find a seat to suit them, or if not, the possibility existed on the next bus, or at least the opportunity, or the belief in the opportunity, was there. Without such assumed abundance, however, the system was exposed as cruel and unfair. Talk of rationing remains a political kiss of death for American politicians. We don't need to divide the pie because the pie will continue to grow faster than the demands made upon it.[7]

Yet rationing is exactly the situation for nature reserves and their management in the oak woodlands outside the Ozarks. There is not enough for every species, every purpose, every partisan. There is not enough land,

and not enough resources to manage minutely those patches that do exist. There are more people than seats on the bus. Choosing among bad options, or assigning seats, is not what the American system does well. Wildlands do not reconcile readily with fire gardening or what might in places be better characterized as fire museuming. Collaborations can compensate but not change the fundamentals. That so many people and so many associations are trying so hard speaks nobly to their practical idealism. But for now, with land management, as with oak restoration, the canopy is growing faster than the light can come through.

MISSOURI COMPROMISE

WHEN HIS WESTERING FINALLY took Daniel Boone to the Femme Osage district west of St. Louis, the Missouri territory remained a marcher land, an unsettled locale between frontiers, of which there were several, each seemingly incommensurate, yet crossing one another like a braided stream. One frontier was political, the division between slave state and free. One was environmental; here the eastern woodlands thinned and the western grasslands thickened.[1]

And one was historical, the place where the old trans-Appalachian frontier ended and the trans-Mississippi frontier began, ready to sweep across the wide Missouri to the Pacific. A few long hunters followed the northward trek of Lewis and Clark along the Missouri River to found the Rocky Mountain fur trade; more trended south into the Ozarks, where they recapitulated the world they had known all their lives. In the winter of 1818–19 Henry Schoolcraft traveled through the hills and recorded an account that could have applied to the whole long generation.

> The settlement at Sugar-Loaf Prairie consists at present of four families. . . . These people subsist partly by agriculture, and partly by hunting. They raise corn for bread and for feeding their horses previous to the commencement of long journeys in the woods, but none for exportation. No cabbages, beets, onions, potatoes, turnips, or other garden vegetables are raised. Gardens are unknown. Corn and wild mats, chiefly bear's meat are the staple articles of food. In manners, morals, customs, dress, contempt of labour

and hospitality, the state of society is not essentially different from that which exists among the savages. Schools, religion, and learning are alike unknown. Hunting is the principal, the most honourable and the most profitable employment. . . . Their system of life is, in fact one continued scene of camp-service.[2]

A single generation, Boone's, liberated by the American Revolution, had made a long hunt from the hinterlands of the Atlantic to the western tributaries of the Mississippi.

It was their sheer westering that had sparked these proliferating frontiers. What had been separated, joined. What had found common cause among the British colonies in expelling Britain now split over who would control the West. The progeny of that Great Migration came to rest in Missouri. Those new lands destabilized the old political equilibriums, particularly between slave and free states. The entry of Missouri into statehood nearly stressed the system to the breaking point, and forced an accommodation, the first of several. The Missouri Compromise of 1820 was enacted the year Boone died.

Something similar may follow America's fire revolution that raged from the mid-1960s to the mid-1970s. Insurgent groups had united against a common foe—in this case a hegemonic commitment to suppression. But as new lands became available for absorption into the new order, one committed to fire's restoration, they could quarrel about means and ends. America's fire polity split into two dominant creeds. One looked to wilderness as a guide, and tolerated human activities insofar as they led ultimately to the removal of human presence in favor of fires that could free-range as fully as wolves. The other looked to working landscapes for which fire remained an implement for hunting, herding, logging, and other forms of sustenance that serve human economies. There was little common ground between them: any land, it seems, must ultimately subscribe to one or the other. The lines between those two visions, often with legal and political sanction, are rigidly drawn. This time the national polarities do not align north and south but east and west. The wilderness ideal remains firmly anchored in the public domain of the West; the working landscape, in private ownership for the most part, or on the public lands providing recreational services, in the East, especially the Southeast.

Missouri sits between them, a middle ground—middle geographically, middle thematically, middle politically. It remains fundamentally a landscape of the border, settled when the public domain was being sold off or handed out as quickly as possible ("doing a land office business" was a phrase with literal punch). Over the past few decades this landscape has become again unsettled, a frontier in the environmental contest between the wild and the working. Out of it perhaps is emerging a new Missouri Compromise.

———————

The Ozarks form a modest uplands, spanning southern Missouri and northern Arkansas, grading into foothills eastward along the Ohio River and westward into eastern Oklahoma. Its core is a granitic dome, long ago leveled, and then raised again into a shallow plateau. That uplift entrenched the major rivers, complete with meanders, and it kindled a new era of erosion that dissected the plateau into an intricate lacework of hills and hollows.[3]

They constitute a distinct landscape for fire. Compared to the Great Plains, they broke down the capacity of fire to free-range. The bluffs, the spring-fed streams, the ravines—all fragment the ability of wind-fetched flame to soar untrammeled. Compared to the eastern plains, etched primarily by streams, the stony rims add texture to the terrain, thus doubling the resistance offered to fire's spread. It was possible in the east to amass burned patch by burned patch into extensive prairie peninsulas and barrens, particularly on karstified limestone like the Pennyroyal that seasonally removed streams. But the topographic texture of the Ozarks fractured even those features into smaller parcels, many of which were less readily fired or given to grasses. Early observers thought the biota similar to the prairies and the terrain similar to Appalachian hills. In his 1819–20 journal of travels Henry Schoolcraft described the routine burning of uplands and slopes.[4]

Both biotic realms, western prairie and eastern woodlands, thrived in the Ozarks but in different settings. The rolling uplands were savanna woodlands; the ravines held the thick forest, tucked away from wind-driven flame. Perhaps a third of such woods was shortleaf pine; the rest, a mixed oak-hickory hardwoods. Dry lightning is rare. Fires are set by people, and like people they have to struggle to overcome the tendency to split and diminish any movement through the hills into ever-tinier tributaries, a kind of reverse stream, splintering into rills and springs of fire as the process proceeds deeper into the plateau.

As the entrenched rivers deepened, and then meandered, mesas were sometimes left within oxbows, which further eroded into a still deeper isolation, what became known locally as "lost hills." Geographically and historically, the Ozarks were themselves a lost hills. Geologically, the Ozarks stand as an outlier and muted echo of the southern Appalachians, much as the Black Hills do for the Northern Rockies. Ecologically, they are a triple junction where the oak woodlands, the southern pines, the tallgrass prairies converge.

The Ozarks are not prime farmland, although there are bottomlands that qualify, but its interior was largely shunned by colonizing agricultural-ists. It knew the usual sequence of prehistoric inhabitants, from Archaic to Woodland peoples, before feeling the outer touch of the Mississip-pian civilizations. It lay on the margins of those cultivating civilizations that claimed the humid bottomlands of eastern North America, raising maize and building mounds. While relics remain to testify of these vari-ous occupations, the peoples themselves had gone, perhaps through that mysterious collapse that swept away so many societies across 14th- and 15th-century North America, from the Anasazi to the Hohokam to the Mississippian Oneota. Throughout, the Ozarks were likely occupied sea-sonally, part of an annual cycle of hunting and foraging—a Barrens in the hills. They abounded with game from turkey to bison, deer, and elk. By the time exploring naturalists arrived, and trees in the mid-17th century began recording fire scars, permanent occupants had vanished. Their fires left with them. The Ozarks became a fire sink.[5]

That changed in the early 19th century when the Cherokees, dislocated by the border wars in the southern Appalachians, began to arrive. They found a kindred landscape, well suited to their economies of hunting, forest farming, and foraging, but one they set about fashioning into still more usable forms, for which fire served as a universal catalyst. They were joined by long hunters and their families, who were, by the reckoning of most observers, indistinguishable in their land use from the displaced natives.

The record of burning ticked upward; and when drought overlay the hills, it became widespread. The burning dappled the Ozarks with prairie pockets and barrens, balds and glades, and where the prevailing westerlies could blow freely, as on the uplands, oak savannas emerged of varying purity. Early observers reported that "both the bottoms and the high ground" were "alternately divided into woodlands and prairies," that it was overall "a region of open woods, large areas being almost treeless," and that the prevailing cause of this action was fire, for "it was common practice among Indians and other hunters to set the woods and prairies on fire." Later naturalists like Curtis Marbut concluded that the open character of the scene was "without doubt, wholly or principally due to the annual burning of the grass." Carl Sauer summarized the record by noting that some fires were set in the spring or fall to improve grazing, and thus draw game to preferred sites, that some were set "to drive game toward the hunters," and that such fires were mentioned "by almost every early writer as the cause of the prairies." As domestic livestock replaced wild ungulates the burning persisted.[6]

The record of burning waxed with each surge of immigrants and waned when they decamped. Yet even when thriving, their flames could not propagate everywhere. They constantly ran into ecological baffles and geologic barriers. On that roughened terrain the swells of flame that rolled with the westerlies from the plains broke, like a storm surge against a rocky isle, splashing forward but with spent momentum. Something more would be needed to overcome those internal checks—more people, greater biotic leverage, more firepower. An 1828 treaty sent the Cherokee to Oklahoma. Their forced removal meant a forced eviction of fire. But already a new wave of colonizers was probing into a land partly broken to an agricultural halter before lapsing into fallow. The newcomers preferred to hunt rather than herd, and to herd rather than plant; they had coped with the oak woodlands and barrens since they breached the Appalachians. They were a loose-jointed, restless society that worked best when moving and became troubled when stuck. What spared their settlements from full ruin was the intrinsic dynamic of the frontier. It struck, broke, and moved on, leaving to others the tedious task of gathering up the ecological shards and remaking landscapes into viable habitats.

In Missouri the earliest settlements clung to major riverways, which served as routes of transport and trade. But the broad Missouri River that bisected the state also defined its two biotic realms, the prairie loess to the north and the forested highlands to the south. Vast prairies were not landscapes a backwoods society favored: they were a place for the plow, not the long rifle. The French clung to the rivers; Germans sought out bottomlands and modest hillsides; the Scots-Irish pushed into the interior, where they could hunt, trap, put down maize plots, and loose their herds to fatten on the abundant grassy glades that served as ready-made pastures. In brief, the newcomers favored places akin to those they knew. The floodplains were fever ridden and prone to cholera; the highlands allowed the newcomers to scatter, as though the frontier were temporarily suspended. When asked why they settled the Ozarks rather than the farm-lusher plains, the pioneers said simply they liked the hills. They resembled the frontier they had tracked across.

———

As did those before them, they began claiming the land by remaking it in their own image. They hunted, they gardened, they turned out hogs, goats, horses, and cattle onto the hills as an open range, and they burned. The numbers of fires increased, rising with populations of people, cattle, and hogs. Livestock granted biotic leverage, amplifying the effects of fire and more than replacing the fast-hunted indigenous fauna. Free-range grazing, in particular, invited free-range burning. Soon every hollow and hillside found its match.

Not only fire's numbers but its sites and seasons changed. The Cherokees had preferred setting autumn fires associated with fall hunts on the uplands. This had the added benefit of forcing game to find winter forage in the bottoms and canebrakes, closer to encampments. The newcomers, with livestock to sustain them, mostly burned in the spring, not wishing to strip the uplands of winter forage and pushing for a quick flush of fodder to plump up the stock after a lean winter's fare. This altered fire regime modified the composition and dynamics of the Ozarks landscape.

Still, such ecological nuances were secondary to the sheer increase in numbers of fires and their propagation throughout the countryside.

People and their stock overwhelmed the internal checks that had held fire to grassy patches between bluffs, creeks, and southerly exposures. The rains were good enough to keep growing something, and a fire-catalyzed economy kept the land constantly kindled. The fires filled out every nook and cranny. This repeated firing and quenching tempered the Ozarks into a hardscrabble landscape. The fires worsened as a logging rush after 1890 replaced shortleaf pine with slash, and as oak thickets replaced savanna woodlands, and as more and more of the flora broke down into biotic rubble and rock. Visiting the hills, Aldo Leopold concluded that many people burned simply to shield themselves from all the burns others were setting. The Ozark candle was burning at both its ends.[7]

By the 1920s the Ozarks were a shambles. To Sauer's mind they were less retarded than the Appalachians of eastern Kentucky and Tennessee, and he distinguished within them "between the farmer of the larger valley and the farmer of the hillside or small cove," but the people remained relatively isolated and backward, and their lands were a mirror of their sinking circumstances. He thought that parks, forest reserves, and recreation were among the best options for the future.[8]

Its chronicle of fire again records a decline, this time not because people had left but because they had stayed, and in fact multiplied along with their livestock, for the land could no longer grow enough to support fodder for both slow-combusting herds and fast-combusting flames. That old economy of frontier burning had no new lands to move into. Resprouting oaks took over sites once under pine, feeding hogs in ways that pine roots could not. Pasture degraded. Erosion worsened. The felled forests left a scalped and furrowed dome. Then drought and Depression forced another emigration, and state and federal governments intervened to acquire significant tracts of land—a new, reversed round of treaties, as it were—and they imposed doctrines intended to evict fire from the land. Even as the biota rebounded, fires diminished in number and shriveled in size. The fire history of the Ozarks once more tracked its human history.

Over the past century the Ozarks have experienced another cycle of migration, another reformation of landscapes, and another long-wave cycle of fire. Neither emigration nor immigration is as complete as those of the 19th century, and the emergent landscape is fragmented, with large patches still mired in the old order. But the long rhythm of burning is unmistakable. From 1581 to 1700, the mean fire interval in the southeast Missouri Ozarks was 15.8 years; from 1700 to 1820, 8.9 years; for 1820–1940, 3.7 years; but since 1940 it extends to 715 years. Across some 500 years the landscape for burning had blossomed and then disappeared.[9]

By the 1920s the Ozarks were breaking down, and they crashed during the drought and Depression of the 1930s. Once the orgy of cutting passed, and its slash had burned, fires thinned, and those that survived weakened, due to the sheer accumulation of the human presence. There wasn't enough to burn in the old way. Then people began decamping, lands fell into tax delinquency, and the flinty stubbornness of Ozark political culture cracked. The removal of the human hand created a new frontier, as land changed ownership or acquired a new cover, or both. Missouri came late to the conservation but it came with hard-wrought compromises that bequeathed an institutional steadiness.

Between 1929 and 1933 the General Assembly authorized the federal purchase of forest land under the Clarke-McNary Act. Soon the U.S. Forest Service acquired 1.3 million acres to make the Mark Twain National Forest. The election of 1936 established a Missouri Conservation Commission, later renamed the Missouri Department of Conservation (MDC), which oversaw forestry, fish, and wildlife, and began acquiring lands of its own, apart from state parks. Decade by decade, slower than activists wished, but with a steady if oft-spasmodic tread, the institutional apparatus for state-sponsored conservation matured. In the 1960s the National Park Service entered seriously into the consortium with the Ozarks National Scenic Riverways. In 1976 and again in 1984 voters approved a sales tax devoted to conservation programs. By 2000 over 13 percent of Missouri, capturing all the critical conservation elements, was lodged in protected public landscapes.[10]

Thoughtful observers had long agreed, however, that genuine conservation could only follow from grassroots popular support, not elite control over bureaus. The private sector controlled most of the land and would determine the grand mosaic of Missouri habitats. A spectacular fusion of private ownership and public service commenced when Leo Drey, beginning in 1951, began developing the immense Pioneer Forest. Eventually his holdings grew to 160,000 acres, all dedicated to sustainable forestry through selective cutting and intimate knowledge of its intricate mosaic of sites. That experiment helped establish a pattern, if sometimes grudging, of cooperation between private and public sectors. Subsequently, donated land, private reserves (such as those belonging to the Nature Conservancy), and conservation easements have expanded the realm of rehabilitated hills.[11]

Together they formed a mixed economy of ownership, some private, some state, some federal, but all working landscapes. The national forests housed CCC camps, which set about stabilizing soils, replanting hillsides, and stopping fires. Peck Ranch, in particular, became a showcase; and when oak decline threatened to spread from Arkansas, the MDC was willing to log off 17 million board feet to halt it in its tracks. MDC brought back the turkey, and even exported its thriving flocks to a dozen other states. But the most potent measure was fire control. The MDC made fire protection a foundational program, assuming that ending the biota-stressing flames would allow the land to recover. Pioneer Forest banned burning of all sorts.

In the early years locals often resented the new order: they regarded fire lookouts darkly as prison watchtowers, and told of soaking a rope in kerosene, setting it afire, and dragging it behind a galloping horse through the woods. Federal foresters were so exasperated that they arranged for anthropologists to study the local residents as one might Inuit or Trobriand Islanders. But the tide had turned, and that creaky pioneer culture could no more hold its own ebb in place than Knut could stand against the sea's advance. The last blast of burning came in the spring of 1953 when an insurgent outbreak of fire affected an estimated 80 percent of the Missouri Ozarks. Thereafter, fires stayed on private lands, or if they strayed onto the public estate were quickly rounded up. In 1967 Missouri at last banned open-range grazing. That reduced the primary motive for continued folk burning to vandalism, a kind of flaming graffiti, unmoored from its economic piers. The old regime collapsed.

Yet after the sighs of relief passed, after the land had recovered sufficiently to regrow pine, oak, and a midwestern scrub, another, if predictable, fire problem emerged. There was not enough fire to make the landscape habitable according to the definitions of the society that was reclaiming the Ozarks. Left to themselves the Ozarks would ecologically transform into something people couldn't use and didn't like. The region had been settled on a roughly midlands border pattern, though a footloose piedmont and mountaineer model replicated neither New England village nor coastal-plains slavocracy. The imprint of those origins endured, as did many progeny of the pioneering generation and a peculiar political culture. The rebuilt Ozarks remain a working landscape, not a wild landscape. But "working" has acquired a new definition. It means recreational, not subsistence, hunting; biodiverse habitats, not open-woods pastures and gardens; sustainable logging, not landclearing and long fallowing; exurban visitors, not backwoods pioneers. It means fire practices suitable to such ambitions—not a restored flaming front, rolling over the hills in a wave of settlement, but a patchy rehabilitation in which varied fires catalyzed diverse habitats. It means a hard slog of fire reintroduction, feeling what flame might do in the new order. As everywhere, foresters have resisted, still locked in ancient blood feuds with open-range graziers, land-scalping loggers, and fire-promiscuous ruralites. But over the past 20 years they, too, have converted or retired from the scene.

The emerging Missouri consensus features a mixed economy of land ownership and purposes and a fire regime for the working landscapes of a service economy. These are not the practices of the Wild; nor does that vast corpus of fire science devoted to free-burning flame in the Wild hold much pertinence. These are landscapes with people at their core: people set fires, people determine how fire behaves, people decide what species fire will promote or contain, people carry fire across the political roughness of land ownership and the historical roughness of a new era. A new generation wants fall colors from hickory, sycamore, maple, and the 22 species of oak in the state. They want turkey, and otter, and bear. They want prairie patches high with native grasses and thick with forbs. They want glades not overwhelmed by brush and cedar. They want clean streams for floating—the Ozarks National Scenic Riverways was the country's first such protected complex. They want habitat for the endangered Bachman's sparrow, and as a grail quest, enough restored shortleaf pine to reinstate the red-cockaded woodpecker. There are pressures for wilderness, too, but they are tiny tiles (23,000 acres) in a vaster mosaic.[12]

———

To those who consider expansive wilderness as the paragon of nature protection, the Missouri model will seem slow, flawed, and exhausting. They will want immense public estates, and will long for administration by agency fiat, presidential proclamation, and court rulings. Conservation in

Missouri has proceeded differently, never moving much beyond a close-argued public opinion, which has required that advocates convince the body politic, not a court or an agency chief. This means a more tedious pace, often lagging national headlines, but a more thorough political legitimacy. Elsewhere, what one administration can declare a roadless area, another can delist. But in Missouri, public opinion, not simply an appointed official, has approved the measures, and what the public has granted after long deliberation it is not likely to cede casually. Conservation in Missouri must work through multiple owners, varied ambitions, and a deeply plowed field of politics. But when it comes, as it has, it speaks with democratic authority as tenacious as matted oak roots.

The wilderness ideal conveys a purity not only of nature but of politics. It works best on empty public lands. It seeks to deal only with the administering agency, not with the whole messy muddle of civil-society politics, which may not be trusted in the end to make the right choice. And it demands a science as seemingly pristine as the nature it aspires to, one stripped of human agency. That type of politics won't work in Missouri, nor will its science. Both must begin with the anthropogenic landscape, subsequently modified to suit local tastes, not with eco-utopian visions in which humans have vanished and the torch left with the last of the Oneotas.

Across most of America our fire policies, and environmental controversies generally, continue to polarize between the wild and the working. Abolitionists remain intent on banning people from preserves, while traditionalists are keen to defend a way of life whose time has passed and that can seem antiquarian, or even ethically repugnant, to much of the national citizenry. As the founding conflict spreads into new landscapes, the prospects ripen for a low-grade civil war, as each side pulls the middle ground apart, forcing it to choose one polarity or the other, all of one or all of its rival. This time, the contested frontier lies not to the west but between the two fires of either coast. In the 19th century relentless expansion wedged a social fissure into a political chasm. In the 20th the growth of environmentalist agendas amid a rapidly unsettled landscape threatened to do the same.

The new Missouri Compromise is an idea, an organization, a practice, and a moment of history. As an idea it puts people into fire behavior as propagators, it puts anthropogenic fire into ecosystems as perpetuators, and it puts cultural landscapes into the pantheon of protected places worthy of restored fire. As an organization it demonstrates how a consortium of researchers, landowners, agencies, and nongovernmental organizations might pool resources to create the heft and momentum to put good fire back into the land and how to carve a place at the table for a region not typically invited to the banquet. As a practice it shows how complex is the task, and how little a putative natural order can serve as a guide.

But it may be as a moment of history that the Missouri Compromise may contribute most enduringly. It shows how to create, if not a microrevolution, then at least an insurgency, against the Establishment. It's easy to forget how startling the assertions of oak, fire, and history were when they were announced. In the early 1990s the Ozarks, like the oaks, were an outlier, an interpretive anomaly, a freak of fire history. Its fire record contributed nothing to fire science. It seemed to have no genuine fire history—fire was an annoying feature like ticks and gnats, not a vital one—because no one had looked for it as anything but ecological vandalism. Now the issue is how to do fire, not whether it belongs. Once again, people are carrying fire across the pixels, this time the rough terrain of fire history.

Two centuries ago the inclusion of Missouri into the union forced a political compromise. Today, for the American fire community, the Missouri experience suggests the contours of a compromise about how to expand the frontiers of fire management without splitting the larger premises behind the project. It shows how to reconcile working landscapes with environmental ideals. This time Missouri is not a centrifugal frontier that threatens to pull the competing factions apart but a centripetal middle that shows how the center might hold.

EPILOGUE

The Oak Woodlands Between Two Fires

S TAND TODAY AT Cumberland Gap and see another wave of unsettled pioneering, this one fueled by internal combustion. In fact, if Turner had actually stood where he proposed, he could now note another cavalcade, the trek of industrial civilization—the coal mine, the railroad, the macadam road, the auto, urban sprawl—a new frontier reclaiming a formerly rural America passing by.

Two eras, two fire realms. One realm had been characterized by widespread burning of grasses and oak-hickory savannas. Where the exposed surface limestone was broad, it supported prairie barrens thronging with game. The fires burned where the people were. The other realm was fed by coal. Some was exposed in veins, much remained buried. In places they overlap, with fossil fuel underlying living fuels. Elsewhere they sit side by side. Kentucky has two broad swathes of coal, striping the state north and south, one in the Appalachians and one in the west, with the Pennyroyal barrens in between. Illinois has coal underlying its prairies.

The early pioneers through the gap were the type specimen for the paradox of American pioneering: that it destroys the conditions that make itself possible. They accomplished that task within a long generation of long hunters. The early settlers could look back in astonishment at how the game and grasses had vanished and how the trees had thickened and taken over. The new era set about remaking the scene as thoroughly. It has even begun to feed on itself as fossil fuels are burned to remove whole mountaintops to get at more fossil fuels. Certainly the resort to coal, gas, and oil has removed many of the working fires that had formerly washed over the countryside.

In 1908 the route through the gap was paved as part of a federal demonstration project. By 1926 it was upgraded to Highway 25E. Then it was widened to accommodate the greater traffic and destroyed much of the historic value of the site. By the latter 20th century a sprawl of car-powered strip malls, suburbs, and more highways, was reclaiming the old scene—not just at the gap but the region—as surely as trees and bluegrass had overtaken the Barrens. What poured through the gap spilled across the region. The Nashville Basin was becoming an asphalt Barren. The Pennyroyal was being paved. In 1996, at a cost of $280 million, the Cumberland Gap Tunnel rerouted traffic away from the gap proper, and the old highway had its asphalt removed, and the old dirt road restored. Interestingly this occurred at roughly the same time as efforts to restore open fire.

Early American settlement was a flame-catalyzed economy that had lived off the land, that had used fire to make its plants and creatures accessible. An industrial society burned coal and oil, and turned to non-combustion alternatives such as hydropower and even nuclear power to supplement its electricity; working fires leached away from everyday life. The abolition of the open range reduced the incentives to burn. State forestry tightened its ability to prevent open fire. The changes came late compared to most of the nation; the last fire recorded at Mantle Rock was in 1965; and this may be taken as emblematic of the final gasps of the old rural fire economy. What burning remained was a relic or arson. What foresters had proclaimed would happen—the fast reforestation of unburned and unplowed lands—did happen. It just came a lot faster and with a lot more collateral damages than anyone had imagined.

Industrial societies like to set lands aside as nature preserves—that is the good news. But in removing fire by technological substitution and outright suppression it has made that land uninhabitable by its historic species. It has not only extinguished fire: it has extinguished the old biome. It was relatively simple, technologically, to strip away the asphalt that had paved Highway 25E. It is far trickier to peel back invasive woods and weeds, and to reintroduce flame on a scale that can move restoration beyond boutique burning—the equivalent of living history museums and woodcraft exhibits. No one was going to spend $280 million to create the ecological equivalent of the Cumberland Gap Tunnel. Perhaps, though, agile networks of partisans might accomplish what mighty public works have not.

⎯⎯⎯⎯⎯⎯⎯⎯⎯⎯

The practical issues are daunting. But as challenging has been the intellectual fight to defend fire ecology as a part of cultural landscapes, and cultural landscapes as a part of natural heritage, because it means recognizing humans as the keystone species for fire. It means admitting that it would be impossible to re-create historic fire regimes by leaving the land alone. It means connecting many tiny sites into a larger if distributed mass, and many tiny purposes into a larger justification. People created those historic landscapes; people will have to re-create them. Even if we wished to deny hunting, plowing, grazing, and logging, we would still have to burn. That is the vision before the region's pioneering fire managers as they move toward a future that lies west of the past.

NOTE ON SOURCES

OST OF MY PARTICULAR sources are identified in the notes. Probably the two best syntheses of oak and fire ecology are Patrick H. Brose, Daniel C. Dey, and Thomas A. Waldrop, *The Fire-Oak Literature of Eastern North America: Synthesis and Guidelines*, General Technical Report NRS-135 (U.S. Forest Service, 2014) and *Fire Ecology* 12, no. 2 (August 2016), a special issue devoted to oaks. Two articles by Marc Abrams are recommended as an introduction to the subject: "Fire and the Development of Oak Forests," *BioScience* 42, no. 5 (1992): 346–53, and "Fire and the Ecological History of Oak Forests in the Eastern United States," in Daniel Yaussy, comp., *Proceedings: Workshop on Fire, People, and the Central hardwoods Landscape*, General Technical Report NE-274 (U.S. Forest Service, 2000). See, also, the proceedings of the recurring Midwest Oak Savanna Conferences.

For two informative reviews of the Mid-American frontier, see Malcolm J. Rohrbough, *The Trans-Appalachian Frontier*, and Terry G. Jordan and Matti Kaups, *The American Backwoods Frontier: An Ethnic and Ecological Interpretation*. John Mack Faragher's *Daniel Boone: The Life and Legend of an American Pioneer* completed a reconnaissance. Carl Sauer looms so large in my thinking about the region that I will simply refer interested readers to Michael Williams's delightful biography, *To Pass on a Good Earth: The Life and Work of Carl O. Sauer*.

THE PACIFIC NORTHWEST

A FIRE SURVEY

It goes like this—
forty acres, give or take,
of bedlam. A derangement of land
called clear-cut.

. . . .

I walk this mountainside
groin-deep in carnage,
drooling splatters of fire.
What will happen here?
Something surgical, not precise but
cauterant, perhaps
a benediction.

—MILES WILSON, "SLASH BURNING"[1]

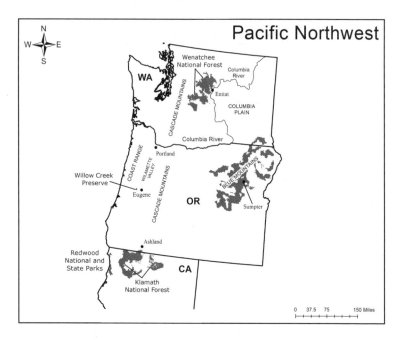

MAP 2 Pacific Northwest.

AUTHOR'S NOTE

Pacific Northwest

W HEN IN 2011 I conceived the notion of corralling stray essays into what became *To the Last Smoke*, I sketched out prospective regions for study. Time and money are the currency of research: I could not treat every possible region, not with the same scope. Initially I designated the Pacific Northwest for a full-body fire scan like that I gave California. But as I weighed the options, I decided that the Pacific Northwest, like the Lake States, was of more historic than contemporary interest—a judgment call that most of the Pacific Northwest fire community will surely protest. I elected instead for a minisurvey similar to that for Alaska and the oak woodlands. The three would complete a volume called *Slopovers*.

By the time I actually began my travels the region had suffered two serious fire seasons, and by the time I ended them it had endured two more. Senators Maria Cantwell and Ron Wyden were making those fires and the general burden of fire protection for the Forest Service into a national political issue. I found I could establish a narrative anchor point, but the continuing conflagrations made it difficult to find closure, which led to doubts that my founding assessment was still valid. Similarly, the Greater Klamath region looked less like an extension of the Pacific Northwest southward than a projection of the California fire scene northward. Regardless, time, money, and publishing arrangements were already committed. The future will decide how much of a right call I managed.

I began my travels in April 2016 with a tour of the Klamath Mountains, mostly Californian, but spilling into Oregon. A family emergency cut short a projected second trip. That took away my time. My money ran out shortly afterward. Over the winter the Joint Fire Science Program stepped into the vacuum with enough funds to complete the task. I made my second trip in July 2017. I wrote the manuscript upon my return. But the fire season, another in a wave train since 2014, quickly overran my conceptual fuelbreaks and pointed to the ways the region was impinging on the national narrative.

Those who assisted are acknowledged in the essays to which they contributed. But let me give them a collective thanks here as well. It's a proud region, full of fire lore and contemporary scholarship and the most serious congressional delegation with an interest in fire issues. I can do no more than sample those wares here. The cognoscenti will recognize all that is missing. Novitiates will, I trust, find enough novelty to encourage them to pursue further. I would be surprised if someone from the region does not complete the survey with the attention it deserves.

PROLOGUE

Green on Black

L IKE MOST PLACES the Pacific Northwest has many terrains and many definitions. And like most, its fire history can be surprising. Among the wettest places in the United States it routinely burns, and has since the ice receded. "Almost every forest type has experienced a fire in the current millennium," James Agee has summarized, "and some may have burned more than a hundred times." Visiting the region in 1912 John Muir concluded that "fire, then, is the great governing agent in forest-distribution and to a great extent also in the conditions of forest growth." Around Puget Sound he concluded that "plainly enough" the forests have been "devastated by fire," though now "veiled with mosses and lichens." So, too, the region's celebrated rains and sodden coasts, and later its environmental controversies, have hidden the ancient record of fire. In nature, in history, in social discourse—green continues to cover black, although in recent years the black has become undeniable.[1]

How to characterize the Pacific Northwest? In geology, it's a series of north–south trending mountain ranges and basins. In climate it shows a similar striping of parallel bands that run from wet west to dry east, the rain-shadow ranges filtering the moisture out. In ecology, the banding translates into temperate rainforest along the Coast Range, montane forests in the central Cascades, and a sage steppe toward the interior.

Old-growth forests harbor marbled murrelet and northern spotted owl; the steppe, the greater sage grouse. There are fires everywhere, sometime.
　　Like most places the Pacific Northwest is a landscape of transitions, as terrain, weather, and woods mutually massage the land into fire regimes, and these regimes align with those of the larger prevailing bioregion. In broad terms, the region divides into wet westside and dry eastside. The westside grasslands, in the Willamette Valley, resemble tallgrass prairie. The eastside grasslands, on the Columbia Plain, behave like shortgrass. The summit of the Cascades divides wet conifers from dry. Southeast Oregon joins the sage steppes of the Great Basin. Northeast Oregon hosts the Blue Mountains, a geologic outlier of the Basin Range and Sierra Nevada and a biotic outlier of the Northern Rockies.
　　The big surprise is the explosive burning, from time to time, perhaps on the order of centuries, of the coastal woods. For that, the region can thank its foehn winds. What the Santa Ana is to Southern California, what the chinook is to Colorado, the east wind is to Oregon. That wet westside, however, has defined the region to the public imagination. Its timbers have concentrated the attention of foresters, public and private. And here green has tended to overlay the memory of black, to hide the scars of big burns, and more recently to overlay a timber industry with a cloth of environmentalist green.
　　The region is no less transitional in the national history of fire. What makes the Pacific Northwest distinctive is the way that fire history plays out against the history of its forests, the timber industry, and public forestry. The fire regime shifted with white settlement from a dialectic of wet and dry to one of fire and axe. Unlike the Lake States, the national timber industry migrated to the Northwest at a time when fire protection was maturing. Unlike the Northern Rockies, ownership was split equally among private, state, and federal agencies, which forced them to form alliances. Unlike California, no master plan linked all the parts together into a common commitment to suppression. In the Pacific Northwest a somewhat domesticated timber industry met a somewhat mature national fire program and found ways to mutually support one another as they coevolved.
　　The region acquired a dense infrastructure of fire institutions, from forestry schools to experimental forests to federal, state, and private fire agencies, all of which demanded cooperation if they were to work

smoothly. From the Pacific Northwest was forged the template for cooperative forestry based on fire protection. Logging slash provided a shared obsession. During the New Deal innovations in fire control spilled out. Later, the region contributed one of the prophets of prescribed fire. The great fire saga of the mid-20th century was the Tillamook Burn cycle, which began in logging slash, then reburned amid massive salvage logging, and ended with a replanted forest destined for commercial harvest.

Call it what you will—character, theme, persuasion—the informing feature of the Pacific Northwest fire story over the past two centuries has been the dialectic between fire and axe. Two wars over the axe and two eras of great fires—collectively, they frame and define most of the 20th century. Then the 21st century kindled a new cycle.

FIRE AND AXE

The First and Second Timber Wars

Axe cuts forest.
Fire burns axe.
Forest covers fire.

THE NATIONAL CHRONICLE of historic conflagrations tracks the spoor of settlement. For most of a century, from the 1825 Miramichi fire to the 1918 fires that blasted through Cloquet and Moose Lake, Minnesota, the big burns feasted on the slash left by landclearing. The North Woods migrated from Maine and New York to the Lake States and then to the Northwest. The outbreaks were most lethal where logging and settlement colluded and rail intensified the slashing and burning. When the Lake States began to falter as its great pineries disappeared into mills and flames, the timber industry looked to the Southeast, where the longleaf would soon be cut away, and to the Northwest, where immense woods had survived the pulse of blowups throughout the 19th century or had grown to maturity in their aftermath.

The Pacific Northwest thus claims a transitional phase in the national narrative between the era of laissez-faire landclearing and that of state-sponsored conservation, and later that phase change within state programs from conservation to preservation. Its 20th century history can be framed by two timber wars. The first timber war hinged on concerns over a timber famine and wildfire. Conservation promised to reduce waste and havoc and ensure a young forest for the future. Fire control would protect existing and growing woods for the axe. The second timber war pivoted over old growth and a fire deficit. Preservation argued to reduce cutting, lessen environmental damages, and ensure the continuance of

old forests. Fire management would work to promote good fires as well
as prevent bad ones.

=========

The first timber war was a classic political brawl of the Progressive Era.
Thanks to cut-out-and-get-out practices, the nation's forests were being
felled far faster than new woods could be found and old ones regrown.
The rallying cry for reformers was "timber famine," which served at the
time as peak oil has for recent generations. Peak timber arrived in 1910
and 1911—Gifford Pinchot declared in 1910 that "the United States has
already crossed the verge of a timber famine so severe that its blighting
effects will be felt by every household in the land." The industry found
it hard to consolidate, and insisted that more cutting was the only way
to stay ahead of the inevitable flames. Conservationists favored reserving
more land in national forests, where the cut could be regulated, or outright nationalization; and they observed that the reckless cutting was the
cause of those terrible fires. Neither side could overcome or silence the
other. But both feared wildfire, and on that one point of agreement they
eventually worked toward a compromise program of cooperative forestry.[1]

   Fire protection was the dynamic weld that bonded state government
with federal, and that soldered them both to industry, either through
special taxes or through private fire protective associations. The first flow-
ering appeared in Idaho in 1906, and quickly jelled after Idaho's Forest
Law a year later that mandated either private protection or a tax for
the state to do it, and so became known as the "Idaho idea." But it was
elsewhere in the Northwest that the idea became firmly institutionalized
into a tripartite condominium of fire partners. In 1909 representatives
from private protection associations in four states (with Weyerhaeuser as
linchpin) met with the Forest Service to discuss fire issues. Out of that
gathering came a consortium, the Pacific Northwest Forest Protection
and Conservation Association, later renamed the Western Forestry and
Conservation Association, that committed to joint programs in fire prevention, public education, legislative lobbying, and mutual assistance. At
its first meeting in 1910 it elected E. T. Allen, a district forester with the
Forest Service, as permanent secretary. What had become a toxic quarrel

among themselves over the axe evolved into a collective fight against their common foe, wildfire. Importantly, fire protection was intended to regulate the axe, not remove it.[2]

On the westside, forests agencies worked to replace bad slash fires with good ones. With little lightning and scant interest in folk burning outside pastures and fields, it was the axe that created the conditions for fire outbreaks. Slovenly slash invited feral fires. The solution was to organize that slash and to mandate burning it; those burns were themselves subject to legal restraints and approved conditions (as measured by relative humidity); and an apparatus was created to attack those fires that escaped. After fires the axe returned to harvest standing boles and cut fuel breaks. Instead of regulating the axe, the interested parties poured their efforts into controlling the fires that too often followed and threatened the future of the industry. It helped that timberlands were roughly divided among public and private ownership.

On eastside forests the problem was different, but the final solution was identical, the elimination of free-burning fire. Here folk burning by settlers, shepherds, and indigenes mingled with abundant lightning to saturate the pineries with mostly surface fires. After logging moved in, these cutover lands, too, became subject to obligatory slash burning. In the working understanding of the agencies, fires that crept through the understory were no different from those that soared through the crowns. Both threatened future forests and invited what officials considered a social tendency toward "lawlessness." With adequate fire protection, however, forests were considered insurable. They could be regrown in the expectation of a future harvest without the risk, as William Greeley once put it, of being swept away on a windy afternoon. A footloose industry could stay and stabilize.

Westside and east, controlling fires could not be segregated from controlling people. Since human engagement with the land pivoted around the timber industry, so, too, was most people's engagement with fire. Fuel management meant treating logging slash; fire management meant burning slash and stopping fires that threatened plantations; smoke management meant handling smoke from pile fires. To a remarkable extent, the project succeeded in its goals. Out of the Siskiyou National Forest came the 40-man crew, progenitor to the interagency hotshot crew. At Winthrop, Washington, appeared the first smokejumper base. From the

Western Fire and Conservation Association came a model for collaborative fire protection, even among blood enemies. The Pacific Northwest became one of the powerhouses of America's fire establishment.

The second timber war boiled over during the 1980s and 1990s. A new environmentalism more concerned with preservation than conservation furnished a social movement and legislative levers. A New Forestry, best articulated by Jerry Franklin, gave it scientific credibility. Snags, windfall, and dead-and-down trunks were no longer just biotic debris, fuel hazards, and lost timber, but legacy structures that helped inform whole forests; fires were not existential threats but an essential process of regeneration. Forestry should seek to maximize all of a forest's assets, not simply its board feet.

The second timber war had several proxies. The first was smoke. Smoke in the valleys was nothing new: residents complained long and hard about it in the 19th century and it was a charge often directed at the transhumant shepherds (mostly Basque) who moved flocks up and down the Cascades. The reward for tolerating a very long, very wet winter was a dry, clear summer, unless smoke saturated the skies. Fortunately most slash burning occurred in winter, when rains could help cleanse the scene.

As cutting accelerated, particularly on public lands, there was more slash to burn, which meant more smoke to linger in the valleys, increasingly converted to a service economy and filled with urbanites, many of whom hated the ugly and damaging clear-cuts, and sought leverage to reduce or eliminate them. If it was bad to breathe second-hand smoke from cigarettes, it must be bad to breathe the wood smoke that seasonally poured into cities and suburbs. Besides, if you controlled smoke, you might control the industry that produced it because without slash burning it would not be possible to replant on logged sites nor to contain the wildfires that would gorge on the combustibles and that threatened the uncut standing stocks and reseeded, highly vulnerable plantations.

The search for alternative ways to manage the mountains of "fuel" left by the axe so obsessed fire agencies throughout the region that it became a national phobia. For most of the 1960s, 1970s, and into the 1980s the problem of wildland fuel meant slash, and most prescribed burns were slash fires. The smoke campaign spread into agricultural fires as well—burning was widespread for grass seed production. Gradually, smoke was strangled out of the scene. The Willamette Valley began shutting down field burning in the 1980s, culminating in legislation that capped the allowable acreage at 40,000 acres (from 180,000), and in 2010 shrank that further to 25,000 with a goal of complete elimination. Spokane followed suit, slashing acreage by 30 percent in 1998, with the goal of abolishing it altogether. Until the digital revolution arrived, the issue of smoke from forestry and agriculture went to the heart of the region's political economy because without slash disposal industrial logging could not succeed. An attempt to haul off "merchantable material" failed because no market existed and the practice damaged soils. So desperate were foresters to rid themselves of slash that one fire officer proposed in *Fire Control Notes* that it simply be buried.

The more powerful proxy was the Endangered Species Act (ESA). The marbled murrelet and especially the northern spotted owl seemed to require old-growth forests to survive, and as those forests were felled, the owl faced extinction. With the ESA as a fulcrum, environmentalists threatened the economic model of the entire industry, which sought to liquidate the stocks of old-growth ("decadent," in the language of foresters) woods before reseeding to plantations. Every value—ecological, aesthetic, even other economic assets—the public might have in those woods was funneled into one output, timber. The ESA was, in fact, nested within many pieces of environmental legislation that challenged the character and consequences of postwar logging on air, water, land, and life. The reliance on large-area clear-cuts as a harvesting strategy threatened them all.

The controversy over the northern spotted owl marked the onset of hostilities that announced the Northwest's Second Timber War. The fighting paused when the 1994 Northwest Forest Plan brokered a cease-fire. The plan caused the timber harvest on federal land to plummet; the era of postwar logging passed a second peak timber. The plan also forced fire agencies to reconsider how to protect those landscapes without appeal to the solutions that had more or less worked over most of the 20th century. Once more, green covered black.

But apart from the Northwest, change had come to fire protection. Simple fire control had metamorphosed into a more pluralistic fire management amid a revolution that sought to restore good fire, and that tended to lump slash burns with bad fires. While industry and the U.S.

Forest Service might rebrand slash fires as "prescribed burns," that linguistic sleight of hand fooled no one seriously invested in the controversy. Instead of sanitizing slash burning, it only tainted broadcast burning for more ecological purposes. And to close the triangle, about the time the Northwest Forest Plan was promulgated, fires began to mutate into more virulent expressions. They burned hotter, they burned bigger, they burned more severely. By the turn of the century the term *megafire* was replacing *blowup* as the stock expression for a big burn. An old plague, thought banished, had reemerged with vehemence.

This time the Pacific Northwest's contribution to the national fire narrative was indirect. It demonstrated the power of the environmental movement, and it helped to disable the Forest Service, still the keystone agency in the national infrastructure for fire. If the Forest Service faltered, so did the country's fire system as a system. The Pacific Northwest had helped propel Forest Service leadership in fire protection. Now it helped unwind it.

There was a sense, among many observers, that this emergent fire plague might bring the sulking, snarling rivals together as the Western Forestry and Conservation Association had a century earlier. Outside the industry the conviction grew that fire had to be managed on a landscape scale, that the only way to contain bad fire was to substitute good fire, and that no one agency could do the job alone. Collaborative forest restoration replaced cooperative forestry. The number of shareholders multiplied to include a civil society of nongovernmental organizations, citizen groups, and nonprofits, along with many federal agencies and interested tribes. But it did not, 20 years after the Northwest Forest Plan, include industry. That left it sated with ideas and starved for funding.

The sticking point remained the trees and how, or if, they should be cut, either before fires and after fires. Bitter controversies remained over how forests should be managed and to what end; considerable public unease persisted with clear-cuts and their ecological (and aesthetic) aftershocks. Instead of fire protection bringing the parties together, the controversies over the axe worsened the prospects for fire management. In many landscapes it was believed that pretreatments were necessary before good fire could be reintroduced, but those "mechanical" treatments could look a lot like logging, or its kissing cousin, thinning, and might require roads, and maybe some kind of forest products industry to help pay the costs. An appeal to "fuels" as a defining metric for fire management looked like silviculture under an assumed name, and labeling as "prescribed fire" what to ordinary folk looked like axe-enabling slash burning only deepened suspicions that foresters were playing a shell-and-pea game. The 1995 Timber Salvage Rider, tacked on to an emergency supplemental appropriations act, which allowed for logging after fires, thus evading the restriction of the Northwest Forest Plan, only further poisoned the association. Instead of fire management calming the quarrel over the axe, the quarrel over the axe threatened to spill over and contaminate fire management.

In this new dispensation the tenets of Progressive Era reform would be challenged as well as the practices and policies it had promulgated. In the first timber war it was assumed that science would inform management, and that experts grounded in that science would apply solutions in—ideally—a disinterested way. In principle, granted enough resources, it would be possible to get ahead of the problem. In 1911 William Greeley asserted that firefighting was as amenable to scientific study as silviculture. As chief forester during the 1920s he made cooperative fire protection a centerpiece of his tenure before decamping to become executive secretary of the West Coast Lumberman's Association (a revealing commentary on the gravitational pull of industry).

In the second timber war scientific research had been fundamental in establishing the requirements of the northern spotted owl, but the protest had followed from a change in cultural values that pitted owls and old growth against board feet, and it was less vital in fire's management. The sense grew that, outside of communities and municipal watersheds and some select biotic sites of high values, we would not get ahead of the problem. Much as a phase change in social values had led to the owl research, so it was likely that fires would lead and the science follow. Fire officers look to managed wildfire to do the heavy lifting and put good fire on the ground. This is what a rational compromise looks like today. One wonders if public skepticism will stir the ashes as hillsides bristling with burned boles replace those with burned stumps.

The first timber war wanted wildfire extinguished, and would tolerate slash burns to dampen the fuels that powered them. It ended with a condominium among the contestants. The second timber war wanted good fire promoted, decided that slash burns were not among them, and argued, if reluctantly, that managed wildfire was the best means to boost burning. New cooperative arrangements, now called collaborations, abound but they lack the binding common cause of their predecessors, not to mention the political and financial sinews. Industry enters the program as a paid service, not as a paying constituent. The second war rests on an ongoing truce rather than a victory.

To track this historic evolution, consider three fires, each roughly 70 years apart, each typical of what a big fire meant in its day.

The 1868 fires raged from the Olympic Peninsula to the California redwoods. Probably over a million acres burned from August through September. Hundreds of small fires—landclearing, camp, and incendiary—swelled into conflagrations as meteorological bellows drove the east wind. Fires burned around Victoria, Seattle, Olympia, Yaquina Bay, and Coos Bay, where a single fire blasted over 125,000–300,000 acres and lent its name to the whole complex in Oregon. Near the mouth of the Columbia a clearing fire escaped, a backfire was lit by a panicky neighbor, the two fires merged under the breath of the east wind, and despite efforts by considerable numbers of volunteers "an advancing line of fire extended from the very edge of the bay to the mountain tops." Smoke smothered valleys, even the long fetch of the Willamette. Navigation slowed.[3]

Yet despite their immensity, they occurred in country not yet fully settled or logged, and the loss of life and property was less than their size suggests. The outbreak occurred 40 years after Jedediah Smith first probed into the country, 25 years after the first great migration to the Willamette Valley, and nine years after statehood. There was no organized effort to fight it, no argument over what to do in its aftermath. It came and went with the east wind. It appeared as another trial to be faced by sturdy pioneers along with floods and grasshoppers. It blew up half a century before the industrial axe bit deeply into the westside forests and redefined Oregon's economy. The only timber war that existed lay within the contest over land generally, between American newcomers and Northwest indigenes. It was part of a cycle of fires that spanned the settlement era.

The 1933 Tillamook Burn came 23 years after the founding of the Western Forestry and Conservation Association established the terms of fire protection as an alliance of necessity and 22 years after the Weeks Act established the terms of federal-state cooperation, later upgraded by the 1924 Clarke-McNary Act. Slash burning was mostly domesticated, disciplined by law and scientific prescriptions, and those overseeing the process even extended bans to the very act of logging when conditions veered into the extreme. On August 14, amid a general woods closure, a company decided to yard one last log, or a careless spark lodged in the piles, or an aggrieved logger set a spite fire—the exact origin remains ambiguous—flame got into logging slash west of Portland, with the east wind shooting through The Dalles like a bullet down a gun barrel. The slash burn blew up. Over the next 24 hours some 10,000 acres an hour burned through the crowns. Smoke blotted out the midday sky. Debris landed on ships 500 miles at sea. The Oregon Department of Forestry mobilized its staff and those of its industrial cooperators. The U.S. Forest Service called up its usual forces, and this time added the massed labor of the newly created Civilian Conservation Corps; this was, in fact, the CCC's first trial by fire. Untouched, the Burn blew to the Pacific, some 330,000 acres in all.

This, however, was only the first iteration of the Tillamook Burn saga. What followed was an expression of the region's dialectic between axe and fire. Industry mustered to begin an immense exercise in salvage logging, while the CCC felled miles of snags for fuelbreaks, and together they cut roads throughout the hills. The Burn reburned in 1939. Another wave of salvage logging followed. It burned again in 1945. More salvage, culling ever fewer trees, but the scene still seemed inexhaustible. Then came the last of the six-year-jinx fires, in 1951. This time mechanized equipment could be brought to bear, and there was less and less for the flames to feed on. Between them fire and axe had gutted a third of a million acres.

Still, this was timber country and a logging economy. All parties viewed the savage Burn as an environmental and economic disaster. Citizens amended the state constitution in order to replant the hills. School children were bussed to help—such was the social consensus on what should be done. With modern fire control growing more muscular, and with better discipline over slash, the Tillamook Forest regrew into one of the epic stories of American forestry. What had begun as a competition between axe and fire to see which would consume the great woods had become an alliance, as both were regulated into the rhythms of a postwar economy. The Tillamook Burn cycle proper lasted 18 years, but the era it embraced spanned another 50.

Even as that new forest was emerging, so was a new service-based economy, an environmental movement, and a fire revolution. Together they wrote a modern analogue to the Tillamook Burn cycle. The first outbreak, the 1987 Silver Complex, a cluster of fires in southwest Oregon, racked up some 200,000 acres. An overview of the regional fire scene noted that it was "the most severe fire in the last 50 years, and one of the two worst in the last 120 years, yet the acreage burned was only 30% of the average acreage historically burned by wildfire in Oregon." Over the course of 140 years landscape burning had shrunk and big fires had become not just rare but anomalous. The simple dichotomies between green and black and between public and private were less useful as explanatory schemes.[4]

In 2002 the Biscuit fire reburned much the same landscape, then spilled out for more. It began from a lightning bust that started five fires on July 13 within the Kalmiopsis Wilderness. Two monster fires—the Rodeo-Chediski fire in Arizona and the Hayman in Colorado, both the largest on record for their respective states—had already sucked in most of the nation's suppression forces, and fires were popping up in Northern California. There were scant resources to spare for something not threatening communities or municipal watersheds; and possible private contractors had not been adequately vetted by the Oregon Department of Forestry, and were not usable. A spate of lightning fires in a legal wilderness gazetted out of a fire-dependent landscape had a weak claim on suppression resources when the nation was at Preparedness Level 5 (the maximum allowed). The fires grew together, and then grew larger, and kept on growing. By August the complex had outgrown the Kalmiopsis Wilderness, the Siskiyou National Forest, and the state of Oregon, as flames crossed into California. Now it commanded national attention and even international, as firefighters were drafted from Mexico and trained fire officers from Canada, New Zealand, and Australia. Some 7,000 firefighters were on lines by mid-August. Burnout operations aroused concern over their severity. Costs went to a ballistic $150 million dollars. So egregious was the expense that the Government Accounting Office (GAO) was asked to investigate. (The GAO found no malfeasance, only

confusion and clumsiness amid perhaps a dose of bad timing.) Environmental controversies worsened when the Forest Service proposed selective salvage logging. In the final reckoning the fire burned a whisker under
half a million acres.[5]

But money was a proxy fight for the old contest between fire and axe.
Those who wanted the land logged argued that the axe would prevent big
fires, or at least not cede such economic losses to flame; the movement
to reserve federal forest lands for wilderness only led to uncontrollable
wild fires. Those who favored fire management as a vehicle to promote
ecological goods and services argued that the interruption of natural, or
at least historic fire regimes, much in the name of protecting commodities
like timber, had caused the buildup of fuels that powered the burn.

Rather than resolving the controversy, the Biscuit fire could be diverted
to argue for each side's fundamental philosophy. It became notorious for
its cost and divisiveness, not for its role in innovation or as the symbol
of a new consensus. It epitomized what was wrong, not how it might be
corrected. It was not a fire that clarified a strategy, it was a fire that condemned the existing system. In 2017 the Chetco Bar fire reburned much
the same landscape once again. The Silver Complex had mutated into
the Chetco Bar megafire. The Tillamook's 6-year jinx had become the
Siskiyou's 15-year curse.[6]

The 1868 fires were barely fought at all. The Tillamook Burn was fought
as hard as agencies could from start to finish, though their counterforces
were puny compared to that of the fire and its stratospheric pyrocumulus.
The Silver, Biscuit, and Chetco Bar fires could call up firepower unimag-
inable to earlier generations, but it was unclear that full-bore suppression
was either possible or desirable. The first ignitions that led to the Biscuit
fire were left to themselves in the Kalmiopsis; then, when they bolted
out of those legal bounds, crews conducted vast firing operations primarily along the Illinois Valley, sending towering columns of smoke that
terrified residents. Helping power those burns was a buildup of fuels, a
quasi-natural slashing caused not by the axe but by the absence of fire.
Probably half the final area was burned within historic ranges of intensity;
many of the most savagely burned sites were the outcome of burnout
operations amid deep drought compounded by unnatural levels of fuels
amid groves of fire-sensitive tan oak that had thickened in the years of fire

protection. If not so clearly anthropogenic in its causes as the Tillamook Burn, the Biscuit fire showed the continuing interaction of nature and culture. This was hardly a natural fire in the sense of burning in historic patterns with historic outcomes. Fire operations notably expanded the fire's final perimeter.

The early arguments were over the fire and its management. Once the flames were controlled, the argument turned to the axe. In 1933 salvage logging began on a heroic scale, remaking the slopes and, with a broad social license, foresters replanted a new forest. Some 69 years later, that project had been inverted. After the Biscuit burn it was proposed to salvage-log burned trees to the order of 67 million board feet, all outside the Kalmiopsis Wilderness and perhaps 1 percent of the total timber affected. By now, however, the axe had become anathema to significant camps of partisans. The Forest Service was taken to court, and protesters tried to block roads. The courts sided with the agencies and the logging proceeded. There was no effort to replant outside the logged sites. There was, in truth, no consensus about what had been done or what to do next. There was no agreement over what the proper theme of the fire should be. A default narrative laid the issue at the feet of climate change.

The cycle of great fires in the 19th-century Northwest had aligned with mythic narratives of settlement and rebuilding. They were great fires not just because they were big but because they provoked their societies to rise to their challenge. The Tillamook Burn cycle repositioned that story line but not its narrative structure or its moral subtext. Such fires were the dark villain that tested the temper of the hero. The newer cycle more resembles the profile of an antihero. The Biscuit fire had left a sour taste. Of the three monster fires of the 2002 season it was the one that sparked no honored legacy. It was a negative exemplar: the fire that showed how fires that had once brought society together were now wedging the pieces apart. It was the fire everyone seemed to wish would go away, and it helps explain how the region could claim national interest without national leadership. Then the Chetco Bar burn arrived, in eerie fugue with a 15-year cycle. It seemed to confirm that these were the fires and these the issues that would define the new millennium for the region unless its fire community could find an alternative fire or some other narrative through which to refract it.

GRACE UNDER FIRE

The Willamette Valley

THE OREGON TRAIL that gripped the national imagination in the early 1840s did not take wagoneering pioneers to the semiarid summit of Steens Mountain or to the sage steppes staggered within the Basin Range or to the lava-encrusted Cascades, much less to the sodden coast. It took them to the Willamette Valley.[1]

The Willamette was a wetter version of California's Central Valley or, more relevant to most of the trekkers, a western echo of the woods-and-prairie landscapes they had known from the old Midwest. It more resembled Independence, Missouri, than anything west of the 100th meridian. They were going to a new place but one that resembled, in fresh biotic idiom, the places they and their ilk had pioneered for a generation. What it lacked was a full complement of grazers, particularly bison, which meant there was more stuff available to burn. Still, it was the West's tallgrass prairie. As with its Midwest cousin, the only way to contain the trees was to cut them or burn them. Shun fire for even a handful of years, and woods would claim the land.

The valley had a fire history to match. That it was prairie, savanna, and burned was obvious even to the earliest Europeans and Americans. Explorers, fur trappers, artists, settlers—all agree that it was open landscape, sometimes mostly wet, sometimes mostly dry, with woody copses here and there, and that it was open because it was routinely fired. The historic record begins in earnest when in 1825 Hudson's Bay Company began sending overland trapping parties to California, which forced them to traverse the Willamette. "Most parts of the country burned," David Douglas noted. The burns were extensive enough to cause forage problems for the parties' horses, an annual annoyance. The 1841 Wilkes Expedition observed the same. So, too, did settlers. "We did not yet know that the Indians were wont to baptise the whole country with fire at the close of every summer," wrote Jesse Applegate, a settler in 1843—but "very soon" learned. The soil that made the Willamette attractive to farmers made it a fabulous place to grow trees. They had to be felled or fired or the open prairie would become a woods.[2]

No place has a record of indigenous burning sufficient to the wishes of researchers, but the Willamette Valley offers one of the most robust reconstructions. It displays a typical aboriginal fire economy—many places burned, each place fired for particular reasons and at specific times, with fire as both a broad-spectrum catalyst and a remarkably precise ecological instrument.[3]

The Kalapuya occupied the valley. An annual almanac has them gathering in prairies in late spring and summer, while the winter wet season still lingered. Here they collected tubers such as camas and harvested waterfowl. In July and August, the weather began to dry, people moved out of the lowlands, and, as conditions permitted, they did some patch-burning after harvesting grass seeds, sunflower seeds, hazelnuts, and berries, and to promote the growth of hazel twigs used in baskets. At this time, too, the camas plots had dried and were fired. The scope of burning widened as the seasonal dryness permitted—part of a cleaning up of debris and reopening of corridors. By late summer, the higher prairies were being burned to help collect insects like grasshoppers and before gathering tarweed. In October, following an acorn harvest, the oak savannas—the most extensive of Willamette landscapes—were fired. The last phase was to burn along the valley edges as part of communal drives for deer. These specific tasks were supplemented with fire littering, as it were, and fires wended into both mountain flanks through well-worn routes of travel. Though not mentioned in historic accounts, it is likely that parts of the marshlands suitable for waterfowl were also fired outside the rhythms of nesting. The range of plants and animals, a pyric herbarium, is extensive, and so are the calibrated practices associated with them—tarweed, camas, wapato, yampa, wild onion, sunflower, bracken and lupine rhizomes, strawberries, huckleberries, blackcaps, thimbleberries, salmonberries, hazelnuts, acorns, plus assorted fauna from grasshoppers and crickets to yellow jacket larvae, ground squirrels, and whitetail deer.

By the time the seasons had cycled through a year, little was not burned. But much of that burning had been highly specific to place and purpose. Burning to stimulate berries and twigs for baskets was localized. Burning for tarweed occurred just prior to harvest, a light fire that seared away the unpleasant "tar" but left the pods, easing the harvest of seeds. The indigenes burned hazel on a 3-year cycle to encourage withes for baskets, and on a 10-year cycle for nuts. Burning oak savannas prior to harvest removed the wormy and rotten acorns, the first to fall, sanitizing the ground. Some of the burning was pyric horticulture. The game surrounds, which then bequeathed prime browse habitat, were a kind of pyric husbandry. They are, in truth, of a piece with aboriginal burning around the world.

There is little controversy about the Willamette Valley as a fire-maintained, anthropogenic landscape. Speculation concentrates on what happened along the forested edges—how far fire might have seeped into the woods and along routes of travel, acting like fuses. That much of the landscape burning occurred in the fall meant it coincided with the east winds that have powered most of the historic fires along the Coast Range. Surely, there were years in which drought and wind drove flames beyond the hunting grounds and deep into the hills.

And there is another paradox in the mountains. The oldest trees in the Oregon Coast Range are younger than similar trees elsewhere along the coast, and they seem to date to the 17th century when widespread forestation occurred; the majority seems to have occurred in the 19th century. The oddity has given rise to a controversy over how much of that big forest was the result of presettlement conflagrations and how much may have resulted from a cessation of aboriginal burning. Did that old forest grow like those that followed settlement fires, or was it a byproduct of major changes in human use, particularly a removal of burners, as the forests of Amazonia appear to be? Or to pursue the query further, are the monster fires chronicled in the decades after settlement simply fires of the old sort, with some accelerants added from landclearing slash, that now had someone to record them in writing? Or were they artifacts of settlement, burning through a dense understory of conifer reproduction that had thickened with the cessation of traditional burning? The burn scars of those historic fires seem to mimic the ghostly landscapes fire cultivated by the indigenes, with giant wildfires burning in great gulps what had previously burned in smaller patches over space and time. Fires propagate. The fires of the Willamette Valley may have spread ecologically through time, like distant thunder, far beyond the rolling prairies.[4]

What is clear is that the Kalapuya imploded. Several waves of disease reached them well before the Oregon Trail became a national highway, and then malaria arrived. The population plummeted from an estimated 15,000 around the time Lewis and Clark trekked to Fort Clatsop to 600 by 1841. In 1855 the fragments of the Kalapuya, along with scraps of Umpqua and Takelma, were relocated to the Grand Ronde Reservation near Willamina. Their link to the land that had defined who they were was broken.

The newcomers didn't like the fire practices they encountered. They related to the land in different ways and wanted other fires, or often, no fires at all. The conflict began in Hudson's Bay Company days when trapping parties struggled in the late fall to find sufficient forage for their horses. The Wilkes Expedition's scientist, Henry Brackenridge, fretted at collecting plant specimens amid the "burnt and parched" landscape and could not locate the North Star for astronomical siting amid the smoke (he literally could not find where he was). The Kalapuya could use fire widely because they ranged seasonally among the communal abundance of the valley, a plentitude they had made possible; but the newcomers understood land as something owned and fixed, and so had to contain fire to patented plots and times suitable for their introduced flora and fauna; and they established larger zones, fire protectorates, around their scattered holdings. Some burning persisted among grass seed growers, pastoralists, and farmers firing stubble and fallow, but even those flames died out as protests rose against smoke, as oak savannas were converted to vineyards, and growers turned from fire to herbicides and tillage. More and more of the valley saw fire seep away. The big fires moved to the mountains where, like the land, they became feral.[5]

Today, little remains of the presettlement prairie. Of an estimated two-million-plus acres at the time of contact, perhaps 10,000 acres, or roughly one half of 1 percent, endures in something like its historic condition. The loss is comparable to that in prime farmland in the Midwest. Worse, what remains forms an archipelago of sites. What happened to the indigenes happened to their land. Yet against the odds small groups and coalitions are working to maintain what remains and restore some of what has been lost.

The Nature Conservancy's Willow Creek Preserve on prairie, woodland, and forest, located partially within the city limits of Eugene, is a good example. It was created first for urban wetlands, then acquired new partners and purposes like the federally listed Fender's blue butterfly. Across the street was a former demolition derby arena now under Bureau of Land Management (BLM) administration—the good news being that it was never plowed and could serve as a habitat corridor. TNC arranged lease agreements with private landowners in 1981; during the 1990s, it expanded through purchase; and in 1995 the Bonneville Power Administration purchased conservation easements as part of a habitat mitigation program. Today the preserve consists of 519 acres, part of some 3,000 acres in the West Eugene Wetlands, itself part of the 24,000-acre Rivers to Ridges partnership in southern Lane County.

Experimental prescribed fires commenced in 1982, but a program really got underway in 1986. The conservancy lacks the staff to do it alone, but excels in arranging partnerships, and it found like-minded folk in several key agencies. Willow Creek Preserve joined the Rivers to Ridges Partnership, a subregional complex of 16 organizations committed to regional planning and restoration, that includes TNC, the Army Corps of Engineers, the City of Eugene, the Bureau of Land Management, the Fish and Wildlife Service, the Confederated Tribes of Grand Ronde, the Natural Resource Conservation Service, the Oregon Department of Forestry, and the Lane Regional Air Protection Agency. Of the total number eight currently participate in the ecological burn program. Other agencies frequently send crews for certification training.

All this helped solve the capacity problem—the simple ability to muster the resources needed to do the burning. The cultural problem by which a consortium creates the know-how and will to burn and sustains the burning year after year demanded uncommon drafts of social capital. That often comes down to personalities, not just agency character but individuals prepared to invest energy, skills, and patience in a program based on maintenance rather than novelty and that demands a temperament capable of working across decades. The southern Willamette Valley found them.[6]

It's all ecological burning—fuels reduction is a secondary consideration when the primary fuels regrow annually. Instead of burning to reduce fuels and let the ecology sort itself out, the Rivers to Ridges Ecological Burn Program burns to advance habitat and lets the fuels adjust accordingly. On a good year the Willow Creek Preserve burns 100 acres. It tries to burn on a three-year rotation. The Rivers to Ridges program burns 300–500 acres annually.

Challenging capacity in its claim for attention is air quality. Smoke regulation went into force the same time the preserve inaugurated its burn program. Slash burning in the fall and winter became an environmental cause célèbre; more relevant was the persistent agricultural burning, principally for grass seed, in late August and September. In 1988 smoke led to a pileup on I-5 that left seven dead. The allowable land burned began to shrink. In 1991 the Oregon Legislature shrank the acceptable acreage for field burning from 180,000 acres to 40,000. In 2010 a state law capped the maximum acres at 25,000, with an ultimate goal of shutting it down altogether. When you have to burn within city limits, within stringent air quality constraints, with a staff that depends on help from neighboring agencies, amid a population that regards trees, not prairie, as natural, with burn bans always a possibility since the season for prescribed fire is the same as that for wildfire, it isn't easy to meet your own goals.[7]

What results isn't the old landscape, or the old burning pattern, but it is a modern simulacrum that preserves what would otherwise be lost. Not least it creates a culture of fire, and a fire culture is what makes a fire program work. Part of what intrigues about the Willamette Valley story is that restoring lost fires and restoring a lost fire culture have fused.

———————————

Restoring land, restoring how to live on the land—the two could not be segregated.

At the time Willow Creek Preserve was coming into being the Kalapuya were a "terminated" people, the 1954 Western Oregon Termination Act having severed the federal government's formal recognition of the tribe. They were one of 27 different tribes crowded onto the Grand Ronde

Reservation. They were a community, but a people without a reservation or land to call their own, part of a 20th century that has been called an age of refugees. The Grand Ronde community was a jumble of lore, languages, and practices. That includes knowledge about fire. After three generations, segregated from the places and rhythms that had defined the old fire regimes, living knowledge had atrophied to a handful of disjointed practices such as burning out a drainage after hunting. Their fire culture was as fractured as their ancestral lands. Their homeland had been in the prairies and savannas; their reservation was in forest. They would have to rebuild a facsimile of their ancestral land if they wanted a facsimile of their ancestral culture.

Restoration began with official recognition in 1983. In the mid-1990s they opened a casino, which brought in money that they used to develop government institutions and provide for their revitalization and persistence. In time, they organized a fire program—bought an engine, trained a crew, began burning, first in slash, then in prairie. Through the Bureau of Indian Affairs they became part of the federal fire force, hiring out to work wildfires, partnering with the Fish and Wildlife Service to burn on oak savannas and prairies in refuges. They learned by doing—that's how they would fill in the gaps in inherited fire lore; practice would show what you had to do to create and sustain the old landscapes. They joined the fire co-ops in the Valley. The money they earned they plowed back into the fire program. By 2017 the fire program had six Type 6 engines, one Type 3 engine, a tactical tender, and a 20-person initial-attack hand crew, all trained to National Wildfire Coordinating Group standards. The program red-carded 50 people annually and earned $100,000 or more a year. As important, it was generating fire stories and reacquiring the knowledge to manage the tribe's patchwork of lands. They were using the existing fire establishment to get the training and resources they needed to build capacity, both for the slash burning they did on tribal timber lands and the cultural burning on lands they were acquiring to restore prairie and savanna.

Unsurprisingly, Grand Ronde was among the most reliable partners at Willow Creek Preserve. It was not just that both projects shared a realization that the land needed fire, or that restoration was occurring both ecologically and culturally, but that restoration was happening simultaneously in parallel worlds, as it were—a shared land but seen through two visions. Two traditions could perceive those same events through very different weltanschauungs. The critical point was not that everyone accepted the same reasons or viewed events through a single cultural prism that all converted to a common creed, but that they were able to pool interests toward getting fire on the land in ways that restored a working semblance to the presettlement landscape that they valued for different reasons. What mattered on the ground was the doing.

An archipelago of sites, an archipelago of peoples. It seems especially right that fire—humanity's species monopoly, a practice common to all peoples and uniquely to people, a power that joins people to land like no other—should unite the restoration efforts in the Willamette Valley. Pyrodiversity is what helps translate cultural diversity into biodiversity.

Still, it can seem a mash-up.

Paradoxes abound, at times seeming to slide into a parody of postmodernism. Tribes might have to relearn lost fire practices by fusing fragments of oral tradition with written accounts from pioneers, a strange scrambling of notions of cultural appropriation and knowledge that had to be confirmed by practice in the field. But then scientists had to learn that what worked as first principles in a lab might not work in a camas field; they, too, had to learn by doing. Fire suppression helped pay for fire restoration, like mining companies forced to rehabilitate disturbed landscapes. Grand Ronde burned for tarweed and camas; the Nature Conservancy, the Bureau of Land Management, and the Fish and Wildlife Service for the federally listed Fender's blue butterfly and *Lomatium bradshawii*; local fire departments, for training.

Perhaps most aptly, it can be seen as an exercise in American pragmatism, not least because part of what is being restored is that national philosophy, seemingly lost over the past few decades, that spoke to how it is that many peoples might act together, that beliefs matter less than deeds, that experimentation is the way of the world. What could appear to some observers as a lack of legibility was actually a badge of success. What mattered was not shared belief about what the land should be or shared understanding of why they should act but a shared practice that put good fire on the land.

The restored landscapes of the Willamette are not large enough to move the valley or reposition major fire institutions. But they can help connect the dots that not only join habitat to habitat but that meld sites and landscapes to cultural meaning. This is true everywhere, but in most places that recipe is overwhelmed by cultural clutter and landscape scraps. In the Willamette the fusion is—for anyone with open eyes and heart—impossible to ignore. Its practitioners are exercising a kind of artistry that is, to paraphrase Norman Maclean on fly fishing, a kind of grace. Those sites grant meaning and purpose, even if no one relies any more on camas for sustenance and if visitors regard the Fender's blue butterfly as a kind of museum piece. They provide shrines in a society that is otherwise committed to stadiums and shopping malls. Their scatter of burning is lighting candles in the dark.

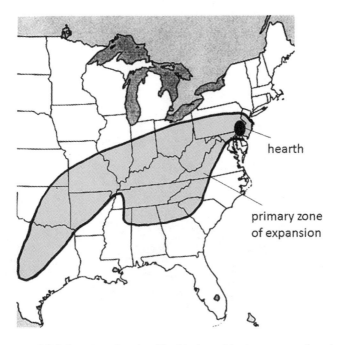

FIGURE 1 Mid-American frontier. The black oval is the apparent hearth for the culture of backwoods settlement. The grey area is the primary zone of the frontier as identified by the prevailing economy, style of cabins, and other artifacts and identifiers. Redrawn and modified from Terry Jordan and Matti Kaups, The American Backwoods Frontier.

FIGURE 2 The Ozark Plateau, dissected into valleys and ridges.

FIGURE 3 Willow Creek Preserve, outside Eugene, Oregon.

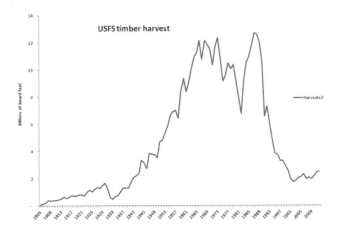

FIGURE 4 U.S. Forest Service timber harvesting. The collapse was delayed in the 1980s, but with or without the northern spotted owl, the cut was too large to sustain. Data from U.S. Forest Service.

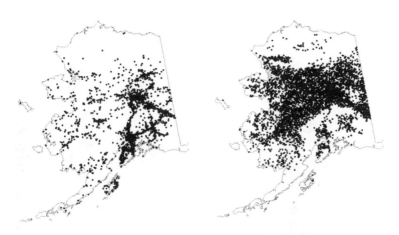

FIGURE 5 Alaskan fire ignitions (1940–2016): (left) from humans; (right) from lightning. Maps from Alaska Fire Service.

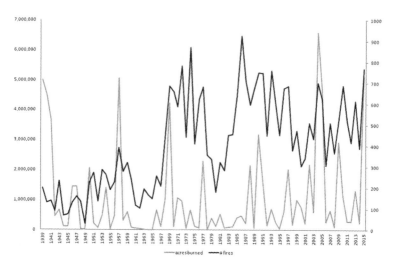

FIGURE 6 Alaska fire history, a chronicle of ignitions and burned area. The plateau rise in ignitions reflects both more people and better surveillance. The burned area varies in typical boreal fashion. Data from Alaska Fire Service.

FIGURE 7 The Aggie Creek fire (2015). Alaska's two fuels, and fires, rub against each other as wildfire meets the Trans-Alaska Pipeline. Photo by Philip Spor; courtesy Alaska Department of Forestry.

CROSSING THE KLAMATH

THE KLAMATH MOUNTAINS, a complex of several ranges, fit nowhere easily in the region because they hold bits of everywhere. They are a triple point in geology, climate, and biota and probably in fire. A transect across them from the Coast Range to the Rogue River Valley provides an interesting cross section of the regional fire scene.

Geologically, the Klamaths mark where the crumpled mélange of the Coast Ranges, Cascade volcanics, and a stray granitic bubble of the Sierra Nevada meet. Climatically, they are the center of a Venn diagram that brings together the winter-rain, summer-fog mediterranean climate of the south, the wet temperate regime of the north, and the semiarid steppe of the east. A great, gouged bubble of stone, the Klamaths became an island refugia as climatic tides rose and fell around them. Samples from all around seemed to make their way into the Klamath ark and, finding niches in its craggy surface, have persisted. The understory contains over 300 species. The conifer forest holds over 30 species. The mountains are recognized as a world center for plant diversity.[1]

As everywhere, its fires reflect its diversity of terrain, flora, and weather. Yet wherever its fire history has been reconstructed in detail, the resulting regimes conform less to the contours of climate and biota than to the practices of culture. What organized that variety into a prevailing order was the patterning of anthropogenic fire. Likely human ignitions laid down a matrix, particularly in the lower elevations, to which lightning claimed the unburned patches. The record of frequent fire is everywhere.

Until recent centuries that ordering was subtle; it tweaked rather than torqued. Only when introduced livestock, hydraulic mining, and especially logging crashed through the landscape did the human hand become dominant either by heaping up fuels, removing surface fire, or imposing slash burns. Humanity's influence went from massaging to mauling. Responsible observers called on fire suppression to intervene.

With the fire revolution of the 1960s and 1970s, the relationship between axe and fire shifted, defined by three major types of landscape. One concerned old growth, which society mostly sought to preserve by staying both fire and axe. A second focused on second growth, whether as plantation or natural regeneration. Here axe and fire had to work together in ways not easily reconciled. The third looked toward grasslands, particularly valleys and relic balds that oaks and tan oaks helped shape into savannas. Here the axe had no interest—these were not commercial woodlands—and managers looked to fire to maintain the grasses against woody encroachment. Each of these landscapes required its own institutional arrangement.

REDWOOD NATIONAL PARK

Redwoods are a California celebrity, unique to California and for *Sequoia sempervirens*, uniquely coastal. They reside within the Coast Range, fringing the Klamaths. They push inland, but not beyond the reach of the fog belt. Their shaded floors, coastal climate, and summer fog—the green through the gloom—all make fire seem alien. Intuitively, the redwood should survive through the ages because it exists in a fire-free environment, and what fires might occur it shrugs off as it would an ant.[2]

Yet the reality is otherwise. Everywhere are the scorch marks, the deep scarring, and the surface scouring of past fires. The redwoods have lived always with fire. They thrive in a fire habitat, routinely burned by light fires, occasionally by intense ones. The historic record, first gathered in the early decades of the 20th century, had landclearing and slash burns in the background. The last major blowup, the Comptche fire, occurred in 1931. Fire suppression arrived shortly afterward. Most observers reckoned that redwoods could survive abusive fires, which in any event came rarely. More recent investigations of fire scars suggest an 11-to-35-year fire return interval for surface burns. Still more recent inquiries, by lowering the sampling plane on trunks, found yet more fires, perhaps on a 6-to-20-year fire return interval. Sampling a nicely scarred redwood at one meter above the surface Steven Norman found the record of one fire; sampling at 10 cm, he found 11. Fires, it seems, gnawed at surface litter like tiny mice. They also burned out of sync with topographic or east–west moisture gradients.[3]

The intriguing biological question is whether the redwoods merely tolerate those fires, or whether they require them to live to the dizzying heights and great ages that are their signature. Unlike Douglas fir, which grows in patchy cohorts, the signature of blowout burns, coastal redwood tends to be even aged, suggesting that fire is more pruner than planter. The complicating management question is that virtually all those fires were set by people.

How do redwoods relate to fire?

Begin with fire behavior. Its tiny needles should compact into dense duff rather than fluffy fuel arrays, its fast growth will quickly put young redwoods above the zone of light burns, and its famously thick bark should render it impervious to most any flame. But redwoods don't shed individual needles: they drop branches that act more like fine slash. Their bark scorches because it burns: its fuzzy surface can carry flame rapidly upward toward the crown where it can sometimes meet deposits of organic duff stuffed on immense branches, there to smolder for weeks, and sometimes to propagate by spotting from treetop to treetop. It's possible that fires, once lodged in the canopy, might linger for much of a season quite independent of anything happening on the surface. The redwoods burn. They just burn in their own idiosyncratic ways.

Its fire ecology is opaque because it's mostly unknown. A few facts belong with basic natural history. Mature trees can scorch without effect, and many burn out cavities at the base, some of which can gradually core through the trunk. The tree survives. A complicating factor are bears that like to rip off patches of bark to reach the sweet cambium underneath, and so injure trees that they can be susceptible to insects and flames (some estimates put the damaged trees at 10 percent or more of the total population). There is no evidence that *S. sempervirens* requires fire to regenerate. Unlike its interior cousin, *Sequoiadendron giganteum*, it lacks serotinous cones and its seeds do not demand mineral soil. Where fire helps the redwood is by hindering its rivals. Regular surface fire inhibits seedlings of Douglas fir and tan oak, and the occasional hot fire can kill off more mature trees of those competitors. The evidence here is suggestive rather than conclusive because it is difficult to test. No one is willing to subject the giants to a fiery experiment directly.

What is truly unknown, however, is the larger ecosystem, and how frequent surface fires shaped it, and what the presence (or absence) of routine burns might mean to the long-term health of the redwoods. Is friendly fire just a common, boring disturbance, with no more effect than sweeping dust off a porch, or does it have some catalytic consequences that fire's exclusion in recent decades will make apparent in the years to come?

So distinctive are redwoods in their coastal setting that it would seem unlikely that fire would figure in their management, and no less odd that Redwood National Park, dedicated to their protection, should serve as synecdoche for the Klamath Mountains, or for that matter the Pacific Northwest. Yet fire long predates the sequoias and their evolutionary ancestors, and today's remnants are full of scorching and fire scars. So, too, the park confronts the same conundrums that bedevil fire management around it.

Redwood National Park is a curious conglomeration, if not an administrative chimera. Protection began in 1917 with the creation of Save the Redwoods League, which pursued a strategy of purchasing critical groves; the state of California matched funds, and transformed the acquired lands into state parks. The great groves were treated as precious art: the parks were museums. That thinking changed during the 1960s as parks were urged to become more natural, habitats were conceived in terms of natural units like watersheds, and species were redefined in terms of population genetics. By now logging had removed 90 percent of the two million acres that had originally held coastal redwoods, and it continued aggressively around the protected groves, threatening watersheds. In 1968 Congress enacted legislation to create a Redwood National Park, adding 58,000 acres under the jurisdiction of the National Park Service. Another 48,000 acres was added in 1978, of which 39,000 had been freshly logged. In 1994 the National Park Service and California Department of Parks and Recreation agreed to jointly manage the collective entity.

This administrative evolution has left patches of redwoods and bordering biomes. The parks' cores are their groves of charismatic megaflora. The additions over the past 50 years include second growth, much of it on land cut over just before the park acquired it. That expansion also pushed inland enough to include critical watersheds, but the ridgelines along those borders replace the redwoods with grassy balds and tan oak. Fire management must deal with three agencies, and three environments, along a strung-out coastal strip, an ecological and political landscape of edges. And it must confront an issue most often side-stepped by the natural allies of preservationists, the problem of a human presence. As with tallgrass prairie, the redwoods grew up within a habitat of abundant, frequent fires almost wholly set by people.

Each of those habitats has its own fire strategy. For old-growth groves, it's fire exclusion. Wildfires are fought, prescribed fires remain unlit. Many of these sites are rightly regarded as irreplaceable objets d'art; tampering with them is too risky either biologically or politically. That decision defers to the future what to make of the long history of fast-recurring burns. At some point fire ecology may have to intervene into fixed-site preservation. But the long fight to save the redwoods was a struggle to protect old growth against axe and fire. Until it is shown that redwoods not merely resist fire but require it, the future points to continued exclusion.

The more intriguing habitat is the large swathes of cutover, now vigorously regenerating into second-growth redwood, along with Douglas fir, tan oak, and a raft of understory species. Here fire does appear to favor redwoods relative to rivals. Judicious surface fires after new sprouts have established themselves, hotter patch-burning as the mixed forest matures—these are plausible fire strategies, and have been tried. The practice balances, in regulated form, a reintroduction of axe and fire. Selective thinning creates a fuel bed, the fires damage conifers and tan oak, the woods open and redwoods can put on a growth spurt. In practice the experiments are small, limited to 20–50 acres. They cost money, time, and effort, and they can be tricky to explain because they seem to violate the purpose of the park, and must be justified as creating the redwood forests of the future. Since that future might be hundreds of years away, a sense of urgency can be distant. The park experiments. No one knows what suite of practices is best.

Paradoxically the most active fire program lies outside the redwoods, and outside the regional competition between axe and fire. The rolling ridges that etch the watersheds were historically grassy, with some tan oak and a few conifers nestled within shady ravines. Without regular burning the conifers infiltrate into the grasslands. Surveys estimate that perhaps 30 percent of former balds have been so encroached that they have become woodlands; so the park burns the extant patches and strips on a roughly five-year cycle. It's a holding operation: the fire program hasn't the wherewithal or urgency to drive back those trees that have encroached and rooted, which would require cutting as well as burning, but they can prevent further losses. The grasslands are a vital habitat and a historic one. Like their geographic setting they are, however, on the margins. Redwood National Park was not established to preserve balds and tan oak.[4]

Redwood National Park is a condominium of state, federal, and private interests within the park as well as along its borders; the character of park expansion has forced a common entity to absorb those tensions internally. All parties have shared concerns, and all have some issues that compete. Only the gravitational pull of the old trees holds them together. What makes the fire scene awkward is the role of humans in the past and what that might mean for future management. That past is blurry, as though viewed through fog. The oldest indigenes have disappeared; the most recent are sequestered onto reservations. The people with no history set the fires with no history. Written records appear at a time of profound disturbance when natives were melting away from disease and conflict, when newcomers loosed livestock over the landscape, when logging churned through the forest. Nor does the redwood help. It grows rapidly but, unlike regions with strong wetting and drying cycles, it doesn't always display crisp rings, easily and unambiguously dated. Still, the abundance of fire scars compensates, and it testifies to a regime that predates the upheaval of contact.

A century of fire records finds few burns caused by lightning, far below what the number and character of recorded fires would require. That leaves humans. People have been on the scene for several times the age of the oldest redwood, which is to say, the great groves grew up within a matrix of routine anthropogenic fire. This seems counterintuitive to many. To natural scientists the fluvial floodplains that seem a preferred habitat for redwoods argues against fire. To social scientists those same sites are equally preferred by people, which argues for burning. Regardless, this is not the familiar scenario in which mindless suppression removed a natural process from a natural setting. And it hints at remediation outside the usual range of treatments.

The question haunting fire ecology thus repeats itself for fire management: were those fires incident to the redwoods or somehow vital to them? If the redwoods thrived despite those fires, then the present policy can be continued without much tinkering. If those fires were necessary—if not to the redwoods themselves, then to the ecosystem within which they grew—then management will have to resort to active burning. Modern fire managers will have to emulate in contemporary context the aboriginal fires that characterized the place for thousands of years.

Each of these management scenes spans the Klamath Mountains, and in fact the region. The uncompromising protection of old growth. The baffling regeneration of landscapes out of second growth. The continuance of grasslands, particularly those of oak and tan oak, both of which were critical food sources for the indigenes. What Redwood National Park does not have is the recent history of large, high-severity burns that have threatened communities, plastered the Klamaths with a kind of ecological shock therapy, and redefined the scale and purposes of restoration. Historically, its dialectic of water and fire shielded the redwoods from such a menace. Today, water does the work, or will unless climate change burns away the fog and allows fires to mutate from the scale of mice to Bigfoot monsters.

WESTERN KLAMATH
RESTORATION PARTNERSHIP

The Western Klamath Restoration Partnership looks like a lot of grass-root programs that want to get good fire back on the land and remove the

conditions that make bad fire a default setting. They bring together representatives of federal, state, local, and tribal fire agencies. They use the Fire
Learning Network mostly facilitated by the Nature Conservancy. They
have adopted the Open Standards Process for Conservation. They look to
Fire Safe Councils for guidelines on structure protection. They appeal to
the Prescribed Fire Training Exchange for implementation. They frame
their goals in language that aligns with the National Cohesive Strategy.
They want to promote fire-adapted communities and fire-resilient landscapes. They want to secure watersheds for towns and fisheries. After long
years of preparation they are putting boots on the ground and torches
into the woods. Their problems are those common everywhere—lack of
capacity and scale.

But they have one element that many others lack. At their founding
they included the Karuk Tribe, which saw the program as a project in
cultural as well as environmental restoration. That meant that traditional
ecological knowledge had to sit next to Western science, that "the revitalization of continual human relationships" with the land was an originating goal, that the essence of the Klamath Mountains was recognized
as a cultural landscape, not a prelapsarian wilderness. To the traditional
circle of claimants, the Western Klamath Restoration Partnership added
another, and not as a token gesture. The Karuk Tribe unsettled the habitual discourse. The contest over future landscapes could not polarize along
the usual lines between the wild and the working (or the wasted) because
the Karuks introduced a competing set of knowledge, norms, and expectations. The scales by which choices might be weighed had three, not
two, platforms.

The history of indigenous fire practices in the greater Northwest, including Northern California, is (perhaps) surprisingly well documented. It
seems undeniable in grasslands, from patch-burned fields of camas to
the Willamette Valley. The record in montane forests, notably pines, oaks,
and Douglas fir is abundant, and furnished the evidence behind light-
burning as "the Indian way" of forest management, and later backed up
Harold Weaver's experiments at the Colville Reservation. It trails off in

the denser Coast Range forests and high peaks. The basic pattern is what characterizes aboriginal fire everywhere: lines of fire along routes of travel and trapping, fields of fire where sites were burned for hunting and foraging along with a degree of fire littering. It was a strategy of fire foraging in which whatever might be burned was.[5]

In the western Klamaths a lot of the burning went to assist basketry, berries, and acorn production from tan oaks and white oaks, as well as for ceremonies, and just cleaning up the land as their duty as biotic citizens. Fall smoke trapped in valleys helped cool streams to boost salmon migration. Around encampments the surface was stripped of fuelwood, establishing fuelbreaks against wildfire. By themselves small populations mean little because people move around the landscape burning by season; they move fire, and fire moves itself. The tribes had a pyric reach far beyond their settled grasp, and their burning was not a decorative applique, used when convenient, but was woven into their hunting, foraging, fishing, rites, their entire social fabric. Western science, ever reductionist, understands fire by measuring fuel, terrain, weather, and climate. The Karuks understand it within a way of life.

That fire culture and the landscapes it created are not easy to recapture or even see. They are hidden from formal history, they complicate natural science and so get stripped from models, and they challenge a wilderness ideology. That traditional knowledge was not recorded because its practitioners were nearly driven to extinction, and because they merged with miners and herders who replaced them, and in some cases intermarried, with a transfer of practices. The Americans who poured into the Klamaths after gold was discovered in 1848 burned even more widely than the tribes they displaced; they just burned for different purposes and in different arrangements. They no longer burned to expose acorns but to expose outcrops. They no longer burned to hunt rabbits and deer but to promote sheep and cattle. Often because of landclearing, their fires— sated with slash—burned wider, hotter, and with less constraint. Both peoples, however, laid down a rude framework within which lightning had to operate, more intricately constrained in lower elevations, looser in higher ranges.

Those set fires persisted longer here than elsewhere in the region— the deeply gouged texture of the Klamaths made access difficult. The

Forest Service did not begin a nominal program of fire protection until 1905, did not begin to reduce burned area until the 1920s, did not begin to extinguish traditional burning until the 1930s, and did not carry fire suppression into the backcountry until the 1940s. In the meantime some burning persisted, either directly through tribal hands or by proxy through ranchers. Traditional knowledge was not wholly suppressed. It continued in a kind of informal economy of burning, an arrangement that gives new meaning to the term *black market*.[6]

It was largely invisible to those who tried to assemble a master fire history. Many researchers tracked the formal fire economy as chronicled in written records and scientific data, noted the waves of burning, inquired of the driver behind them, and concluded it was climate. Compared to planetary dynamics, the cultural practices of relatively small bands of indigenes living on acorns and salmon seemed trivial. In fact, the researchers were asking the wrong question. The right question was not whether nature or culture dominated, but how they interacted. The assumption was that natural conditions, notably lightning fire, determined the matrix within which anthropogenic fire had to operate. More likely, human burning created that matrix, leaving lightning to mop up what humans left or shunned. When climatic conditions favored fire, human firepower increased; equally, human burning added to and reshaped natural fire regimes. The quest to make one or the other a driver looks like a secularization of Judeo-Christian theology in which a jealous Purpose denies all others. The core reality is that fire resembles a driverless car, synthesizing everything around it. In truth, fire was, as the Western Klamath Restoration Plan declares, a relationship.

Now the scene has again inflected. Over the past 20 years large-area, high-severity fires have blasted over much of the Klamaths. The mountains were pulverized by the 2008 fire siege. The usual culprits were identified as legacy fuels and climate. It would seem that in this new dispensation fire regimes do follow fuels and may tack closer to climate change, but that ignores the deeper reality that fuels are out of whack because of human land use and fire practices, notably fire exclusion, and that climate is unmoored again because of human fire practices, notably a shift from burning living landscapes to burning lithic ones. (The point turned lethal in 2008 when eight firefighters died in a helicopter crash, along with the pilot, on a fire in the Trinity Alps Wilderness.)

One of the vital concepts of the Western Klamath Forest Restoration Partnership is that of overlap. The best check against future fire is past fire, and where fires compound, one on another, the landscape becomes more resilient. Even megafires are full of blowout holes and unburned gaps. What matters is getting a sequence of overlying burns along the lines of what had existed historically. It's not enough to reinstate fire: you need a fire regime. A historical reconstruction of how fires had overlapped across the landscape is the basis for planning future treatments.

But before those overlaps can appear on the ground they have to appear in the sustaining society. Groups have to find points of agreement. They have to overlay ends, means, and understandings. Those culturally encoded goals and views don't have to be identical; one group does not need to convert the other. Traditional knowledge and modern science can look at the same scene with radically different perspectives and coexist, with neither the designated driver. But they have to agree on where, when, and how to act. Each vision will run into and over the other and, if done right, build social resilience. The Western Klamath Forest Restoration Partnership shows how to do this, and what it can cost.

ASHLAND FOREST RESILIENCE

The target of the Ashland Forest Resiliency Stewardship Project, itself bonded with the adjacent Ashland Forest All-Lands Restoration, and both broadly nested within the Rogue Basin Cohesive Forest Restoration Strategy, is not a world heritage site, not a national park ablaze with geologic and biological monuments, not a wilderness area, not a recognized hotbed of biodiversity, not even an expansive wildland degraded through mismanaged fire and axe. It's a municipal watershed for the town of Ashland, Oregon. The town lies in the alluvial fan of the Ashland River that splashes down from Mount Ashland on the northeastern flank of the Klamaths. The watershed is more wildland-urban interface (WUI) than wildland, more urban park than nature preserve, but it's a landscape of trees, shrubs, and grasses, and its health is vital to the city.[7]

The most far-sighted of the city's early promoters, men like C. B. Watson, early recognized the significance of stabilizing Ashland's water supply. But such visions were common among Progressives. Watershed protection was a powerful public argument (one with good political optics), and it furnished the constitutional justification for national forests (those waters flowed into navigable streams, thus invoking the ever-elastic interstate commerce clause); and when western leaders—mostly men of commerce—wanted national forests, they urged them in the name of watersheds. In recent decades municipal watersheds have returned as an object of fire management as high-severity fires have threatened the water supplies of cities large and small. Santa Fe and Albuquerque were hit in 2012. Flagstaff suffered flooding and mudslides in the 2010 Schulz fire. Denver was spooked by the Hayman fire, and then suffered the ignominy of having a prescribed burn nominally set to improve its watershed go feral, burn through 133 houses, and kill three people. Both Flagstaff and Denver have recently authorized municipal bonds to be spent for forest restoration, primarily to calm wildfires and buffer watersheds. The three-legged race that is fire and water isn't unique to the Northwest.

Early on, however, Ashland forged a strong nuclear bond between conservationists, local politicians, and business to acquire the necessary land and to see that it was both developed and protected. Such coalitions didn't just self-kindle in the political equivalent of spontaneous combustion. They were cultivated. So it is not surprising that watersheds flowed easily out of the Northwest's first timber wars of the early 20th century. Fire and axe could trash catchment basins, and all parties had an interest in dampening the damages. What is striking is that it survived the second timber wars that ended the century.

Again, this did not emerge of its own accord, as though institutions were spores stored in a seed bank, sprouting when conjured forth. The old alliance was gone. The U.S. Forest Service, long the indispensable agency for cooperative forestry, was wounded and unable to muster the legitimacy needed to rally others around a common cause. It might provide some funds and legislative cover, but it could not, as it had often in the past, forge a coalition around its own mission. Too many potential partners considered it as part of the problem. Instead, a new coalition arose, and though it would work largely on public lands, it came mostly out of civil society. It was a collaboration of partners. No single entity set the agenda.

―――――――――――

The Ashland Forest Resiliency Stewardship Project is the latest in a series of cooperative ventures. The city purchased 700 critical acres. But that was a pittance in a forested watershed that ranged over 15,000 acres and 5,800 feet in elevation. In 1892 Ashland had petitioned the federal government to reserve the slopes of Mount Ashland from further logging and stream damage; the Ashland Forest Reserve was gazetted the next year. That brought regulation to timber cutting and fire protection.

Already the fire scene was morphing. The more researchers study the region, the more frequently they find it was burned. A textured landscape like this will hold lots of fire niches, and support lots of fires, some less common and more intense than others, but coarse or fine, the weave of fire and land testifies to routine surface burns that, as with Redwood National Park, seem to defy gradients of rain and terrain. The ethnographic evidence argues for abundant burning, most of it small, light, adjusted to particulars of years and availability, seasonally migratory, occasionally explosive when flames kindled to burn berries or harvest targrass might catch an east wind. The first scientific party to visit the region, the Wilkes Expedition, moving south from the Willamette Valley, entered the Rogue Valley in September 1841 after days of charred landscapes and smoke and saw an old woman "so busy setting fire to the prairie and mountain ravines that she seemed to disregard us."[8]

But protected forests meant fire protection. The savannas no longer flushed away encroaching conifers and shrubs with fire freshets, the brush no longer burned every few years, the woods no longer showed an open appearance and mixed composition as fire exclusion allowed for a green smear to cover the old black. Around Ashland wildfires broke out in 1901, 1910 (Bushy Creek), and 1959—those that escaped initial attack became more ferocious. For decades fire control could increase its power faster than fuels could accumulate. It helped that most conflagrations needed an east wind, and while Ashland lay in a fire delta beneath a fire floodplain, the worst winds would drive the fire away from the city proper. Then during the 1990s a tipping point was reached, and it became obvious to serious observers that a pyric explosion was possible that could savage both Ashland's watershed and its outskirts. The 2002 East Antelope fire on Grizzly Peak faced the town across the valley. The 2009 Siskiyou fire burned over the hill by the town's south border. At all compass points there are scars from past fires and some not so distant in space or time.

In 2010 the looming threat to life and livelihood prompted concerned parties, including the city, to join a 10-year collaborative effort to restore the watershed to something closer to its historic condition. The enterprise involved a coalition of partners. The U.S. Forest Service contributed $4.6 million, the Natural Resources Conservation Service $3 million, and the Oregon Watershed Enhancement Board $1.5 million—that provided the bulk funding, while Ashland enacted a utility surcharge to the tune of $175,000 annually to supplement. The National Cohesive Strategy for Wildland Fire Management provided target goals. The Nature Conservancy worked through the Fire Learning Network to catalyze the social dynamics. To be effective the project's scope had to involve both private and public lands. To pass muster by skeptics it had to rely on a long gestation to build trust.

The axe was present—some five million board feet of timber have been hauled out by truck and helicopter as part of thinning operations. But the usual dialectic between commercial companies and environmentalists was finessed by having standards set collectively and then arranging for Lomakatsi, a nonprofit organization that specializes in restoration, to oversee the actual work, and then by means of a stewardship agreement, not a contract. (The project has its fudges, like calling roads "temporary restoration pathways.") By the summer of 2016 probably a tenth of the projected 53,000 acres had been treated. A patchwork quilt of prescribed burns dappled the slopes.

All lands, all hands, all values—this is the political formula of modern cooperation. Making Ashland a more fire-adapted community was as much a goal as making its watershed more fire resilient. Restoration would enhance wildlife habitat, improve recreation experiences, and even support a small jobs program. Not everything pointed to fire, nor could prescribed fire alone succeed, but if the Ashland project got fire wrong, the rest would go wrong too. Every added task potentially expanded the coalition of support and reduced the risk to each actor. It also ramped up the investment in effort.

The Ashland Forest Resiliency project has become a model for other communities. This is what it takes to reach consensus, achieve working capacity, tweak the out-of-whack landscape into more habitable forms. But it has also become an exemplar of what is problematic. The scale is too small and too slow to expand much beyond Ashland proper. The project is one tile among hundreds. At some point volunteerism, the American preference, even larded with grant money, is not enough to wrestle with the sum of costs and the scope of needs. The history of America's cities, which used to burn as often as their surrounding landscapes, may be a guide. Those fires did not vanish because of voluntary associations and insurance markets. They went away because laws were passed and enforced to change the character of cities and make them less combustible. Fire protection became mandatory.

What is most impressive about the Ashland project is what is also most damning. It takes a lot of capital—fiscal capital, political capital, social capital. Years of endless talking, engaging, and touring demonstration plots may pass before something happens to scale. This requires a temperament and patience not normal in an age of digital connections, when instant gratification, as the witticism goes, takes too long. So, too, the looming menace of wildfires may not be obvious to a community constantly refreshed by exurban immigrants, and something an economy based on tourism does not care to broadcast. Probably only managed wildfires will allow the project to leap from 53,000 acres to the 4.6 million believed to need treatment, but that is a different model than deliberate, ahead-of-the-catastrophe restoration.

Watching the Ashland Forest Resiliency and its persistent passion is a humbling experience. It should remind us that so much of the current fire scene is about behavior, not the projected paths of wildfires but the dedication and tireless gathering, cajoling, persuading, meeting, chatting, listening, arguing, badgering, meeting again, backslapping, and yet another meeting to review, once more, and then again, those pesky points of difference and resolve the narcissism of small differences. It's less about bold moonshots than the social equivalent of patching potholes and mending fence lines.

Coalition building is more alchemy than algorithms. No National Fire Danger Rating System exists to integrate those factors into an index of fire management. There is no social equivalent to fire forecasting models like BEHAVE or FARSITE. Which is odd, really, because in the end the human dimension is what matters most. We are still the keystone species for fire. We can't control fire's behavior until we control our own. For many practitioners, and even researchers, fire behavior remains the foundational knowledge. But fire behavior isn't something that only characterizes fires. It also applies to the behavior of a fire community.

RESTORATION SINGS THE BLUES

I N SEPTEMBER 1991 the Blue Mountains Natural Resources Institute sponsored an inspection of the Blues' forests. An august interdisciplinary team of scientists toured the woods by land and air, and those "who had not seen the Blue Mountains recently" were "shocked" by the severity and breadth of mortality among its forests. Insects, disease, and fire were reducing the fabled woods to a pittance of their former splendor. Mountain pine beetle hit in the 1970s, spruce budworm in the 1980s, and big fires gorged on the woody carcasses; perhaps 70 percent of the forest was dead and dying. The disruptors showed no signs of diminishing. The Blue Mountains were on their way to becoming the poster child of a national crisis in forest health.[1]

The Blue Mountains inspection team noted that what it termed the "Blue Mountain scenario" was widespread throughout the montane forests of the American West and, in fact, "has been reported in the literature since the 1940s." What changed now was a scale of devastation that seemed "unprecedented," and a broadly cultural concern, reflected in politics, that did not like what it was witnessing, and was perhaps prepared to roll back those noxious outcomes on a scale commensurate with their arrival. "We have taken drastic steps in attempting to exclude fire from fire-dependent ecosystems in the past. Now bold steps musts be taken to effectively manage ecosystems with all processes in place, including prescribed fire."[2]

The alarms echoed off every firewall in the West. *American Forests* announced a "health emergency" in the country's western woods. New Forestry folded unhealthy woods into its critique of commodity-driven Old Forestry. Experiments in southwestern ponderosa pine diagnosed the malaise and prescribed treatments. The U.S. Forest Service, in search of a more environmentally friendly charter, adopted ecosystem management as a new "paradigm." In 1993 the Clinton administration instructed the Forest Service "to develop a scientifically sound and ecosystem-based strategy for management of eastside forests," which gelled into an Eastside Ecosystem Management Project. The 1994 Northwest Forest Plan sought to reconcile environmental values with a timber industry, which would only be possible through a major reduction in cutting old growth. The Forest Service followed with an (unfunded) Western Forest Health Initiative, the first of an escalating series. A variety of causes had led the Blue Mountains into their state of disarray. A variety of responses would try to correct it.[3]

The Blues were where pioneers first encountered the promise of a green Oregon. A century and a half later the Blues were where the consequences of that encounter were first recorded in all their complex ironies.

The Blue Mountains are a cluster of isolated mountains in northeastern Oregon. If the High Plateaus of Utah show a Basin Range structure imposed on the Colorado Plateau, the Blues show a Basin Range structure on the layered basalts of the Columbia Plateau. As the slopes rise, so does rain and snow, which means the forests thicken from low-elevation ponderosa and larch to high-elevation mixed conifer.

All of the Blues' forests burned, though in different regimens. The pine and larch showed the classic profile of a montane forest—burned frequently and lightly. There is plenty of lightning, but undoubtedly the indigenes added their own ignitions. After the Nez Perce and Cayuse acquired horses in vast numbers, another dynamic appeared, as people burned to improve pasture and herds cropped off the grass that fed fires. That dialectic came after diseases were thinning tribes and was less than a century old when Americans arrived in numbers.

It's hard to know what "natural" fire regimes the Blues boasted. What matters is what happened to those that explorers, fur trappers, missionaries, and settlers found. They saw summer fires from lightning and accident; and late summer and fall fires set in fir openings, just before the rainy season, to promote edge, berries, and browse and in ponderosa and larch to keep the surface open and grassy. Inevitably, some fires got big. The late summer, in most years, was sufficiently aflame and smoky to make it "a difficult time to travel through the mountains."[4]

The disruption of the mountains by contact with Europeans began with diseases, with the proliferation of horses, with beaver trapping that unhinged the mountains' hydrology, with market hunting that upset the choreography of fauna and flame, with scattered settlements in the valleys like Grand Ronde, and after rail arrived in the late 1880s, with grazing, logging, and mining, all of which reorganized fuels and fragmented habitats. The old order of fire began breaking down—light fires that had occurred routinely ceased, and large fires that were rare became common.

The economic order depended on rail and mining, which created markets for local herds, timber, and agriculture. That unleashed the big three factors that, throughout the American West, unpicked the warp and woof of prior landscapes. Grazing throughout the mountains stripped away the grasses that made surface fire possible. Transhumance put flocks into high meadows during the summer, and in montane savannas during the winter. Logging quickened, clearing out old-growth ponderosa and larch and leaving Douglas fir and grand fir to flood the gaps. Narrow-gauge rail opened that timber to markets beyond the Blues proper. High-grading the forest—taking the big trees and leaving the now-unburned thickets of reproduction—smashed the structure that had accommodated habitual surface burning. Fire at first flared, then, over time, faded as woods were cut out and their slash burned and as grazing cropped off grasses. Add a fourth, oft-overlooked factor: the removal of native burners. For the indigenes fire was a general catalyst as well as a specific stimulant for hunting, foraging, and horse herding. For the newcomers, awash in slash, it was a threat and its interminable smoke a nuisance. So not only fire's fuels, but its ignitions wobbled, then toppled down.

For decades the encounter with white America had been an exercise in migratory exploitation. Trap out the beaver, kill off the big game, eat out the grass, cut out the big trees, then leave, and start it somewhere new. Eventually there were few places left. By the end of the 19th century, the Blues were an ecological shambles, increasingly as unfit for the newcomers as for those they displaced. What happened in the Blues was happening all across the West. Different rocks, different shrubs and grasses, different assemblages of trees, different watersheds, but the outcome was remarkably similar—the fire histories of the Blues are virtually interchangeable with those in the Cascades, the Northern Rockies, the High Plateaus of Utah, the sky islands of the Great Basin, and the Mogollon Rim of the Southwest.

The state intervened. Even amid the shameless excess of the Gilded Age, the country rallied to protect what remained of its public estate. It was an astonishing decision, as though Americans instinctively recognized that their peculiar brand of capitalism was a toggle switch, not a rheostat. It could not be calibrated, only turned on and off. If on, it tended to push land and society to the brink (and recurringly over the edge). The only way to control such an economy was to turn it off, or in the case of land, fence it off.

Conservation as a doctrine and the national forests as a project sought an elusive middle ground by adjusting, through applied science, what kind of industry the mountains could take. The Forest Service strove to tighten the reins of herders, direct the axe, and generally quell the havoc. That they would oversee the economy through the disinterested decisions of a corps of forest engineers would lessen both the worst extravagance of American capitalism and of the politics it often controlled. In place of greed it would substitute reason; in place of corrupt politics, it would appeal to certified experts; in place of folk wanderlust and superstition, it would offer guided direction and positive knowledge. What is shocking is that a century after observers declared the Blues a mess, after over eight decades of committed research and dedicated conservation through government bureaus, the Blues was once again a mess. What had occurred first through folkways and laissez-faire economics was, it appeared, repeated through official agencies and science-informed forestry.

The Progressive Era sought to bring the intellectual discipline of modern science and the institutional discipline of disinterested bureaucracy to tame the havoc on society and nature. But the wreckage had returned, as though human history followed cycles like beetles and budworms. The new insults often repeated the old ones. The usual suspects continued under new aliases: they just had more idealistic justifications and came through state-sponsored agencies, not robber barons. The primary vehicle was of course the U.S. Forest Service. At the Blues, according to Nancy Langston, it saw "two things: a 'human' landscape in need of being saved because it had been ravaged by companies and the profit motive, and a 'natural' landscape that also needed saving because it was decadent, wasteful, and inefficient." The Little GPs (Gifford Pinchots), as they styled themselves, "using the best possible science," would make "the best possible forests for the best of all possible societies." Instead, they repeated a cycle of "forest chaos."[5]

Much as science had secularized theology, so conservation rationalized frontier economics. But the same commercial structure remained: timber, grazing, mining, some agriculture, towns precariously perched on the rim of a commodity economy. The difference is that those activities were now regulated, at least partially and in principle, by an agency committed to applied science. That agency, however, also brought its own pathologies. Professional forestry had emerged in central Europe as a graft on the great rootstock of European agronomy, and like planting wheat or potatoes, it required the old "wild" or "decadent" woods be first cleared. That is just what foresters set out to do: liquidate the remaining old forest in order to plant a new one engineered to maximize benefits. Aggressive fire exclusion assured the reproduction would come in thick. This strategy was not unique to the Blues: it was practiced wherever possible, not only in the United States but throughout Europe's imperium. In his handbook for Indian foresters, H. H. Champion casually remarked that the "jungle" had to be felled and jungle fires stopped in order to bring rational agriculture to the chaos that was the Indian landscape. The process, full of professional arrogance, continued in the United States into the 1970s until public outrage shut it down.

The outcome only rationalized the previous practice of clearing out the old-growth ponderosa and larch, but this time in collusion with the state instead of collision with it. "In their haste during the 1920s to regulate the Blues forests, planners authorized extremely rapid harvests, well knowing that those harvests would ensure the collapse of the local timber industry by the 1990s." When the economy turned sour during the Depression, the cutting had to continue as well to help stabilize local communities.

Instead of planting and costly silviculture, a future forest would emerge naturally through fire control—fire being declared the implacable enemy of ponderosa, thinning out its reproduction, scarring its trunks, sacrificing the future's needs for today's conveniences. The resort to fire protection as a surrogate for costly silviculture was, again, not unique to the Blues, or to the United States. It was an axiom of global forestry.[6]

What is remarkable about the outcome by the 1990s is how much of it was predicted in the early years. Even in the 1920s forest planners knew the prime timber would be gone in 60 years, and that is exactly what happened—the big timber was cut out, and the promised young timber was either too immature or the wrong species; and because so much of it grew up to grand fir and Douglas fir, not really the right species for that site, they became the primary points for disease and insect infestation. The dense thickets were vulnerable to beetles, budworms, mistletoe, and of course fire. The light-burning controversy that had centered in California early in the century had predicted that fire suppression alone would lead to unhealthy, more fire-prone forests—precisely what happened. Those critics had not factored in high-grade logging, which broke the structure of the forests as well as upending the processes that ran them. It wasn't just axe or fire, but the two, along with grazing and wildlife and everything else, in interaction.

Those early foresters, faced with pandemonium in the mountains, didn't know all they needed—that is the usual explanation offered by the agency. They had not understood the interplay of grazing, wildlife, roads, the full apparatus of settlement and markets on the land. In particular, they hadn't appreciated the longer-term consequences of fire exclusion. The national trauma felt by the Forest Service from the Big Blowup still resonated: the 1910 fires are still their largest on record for the Blues. Had the science been more mature, apologists argue, they could have crafted better policies. Had they tolerated more surface burning, they could have avoided the dismal outcome that by 1991 made the Blue Mountains a byword for trashed forests. That was the official and the guys-on-the-ground take.

But, as Nancy Langston notes, "unfortunately, ignorance was not the cause." Claiming lack of knowledge absolves the Blues (and the Forest Service) of its fire program, as that fire program absolves the agency of the continued cutting. There was plenty of fire knowledge around: it just wasn't codified in forest science. It was in indigenous lore, in folk experience, and often in the evidence of the forest itself. There were plenty of critics, even around the Blues, up to the fire revolution of the 1960s and 1970s, when "the science" flipped and argued for fire restoration. Both sides could appeal to putative "natural" conditions to argue their case. Both could "use Science to silence debate." Both argued on the basis of select facts that supported their position. The Science had, in fact, earlier argued for fire exclusion. It was inadequate, and became politicized, but it will always be inadequate, and if important it will always become politicized.[7]

The scene that shocked that cadre of fire scientists in 1991 had, in fact, been predicted early in the century. The critics had argued empirically but not from institutional science. Some facts, some testimonies, it seems, are better than others.

———

Forest Dreams, Forest Nightmares was published in 1995, but Nancy Langston wrote it, first as a dissertation, then as a book, during the crisis years of the Blues. In it she notes that the 1990 Forest Plan recommended "*doubling* the already inflated harvests of the late 1980s, an increase of five or six times over what most ecologists thought was feasible." The Umatilla Forest Plan proposed that, with good science, grazing could be tripled. "The old dream of reshaping a forest to make it better and better had not disappeared." But it was being redefined. The book concludes as the Forest Service announces its conversion to a paradigm of ecosystem management and launches the Eastside Ecosystem Management Project.[8]

The king is dead, long live the king. The new order looked suspiciously like the old one with more fashionable clothes. The old project would be updated with better technology and a stronger sensibility for other environmental values. "Fire and science, taken together," however, "were suddenly providing managers the justification for something that looked very much like business as usual." The crisis of the second timber wars would be solved much like the crisis of the first: by more active, science-informed management, which for the Pacific Northwest meant the axe. Russell Graham wrote that "nearly all of the effects a fire has on a forest can be accomplished using traditional silvicultural treatments." The solution to the problems caused by technology was more technology. The way to deal with the unexpected consequences of intensive management was with still more intensive management. Langston, however, concluded that "everyone who has ever tried to fix the forests has ended up making them worse." At the core was an alloy of fear and hubris, that "although the Forest Service's goal has changed—sustainable ecosystems rather than commodity timber crops—the basic paradigm of control has not."[9]

The forest health crisis, with potential "catastrophic" fire and unprecedented waves of insects, inspired a wave of programs, from the National Fire Plan to the Healthy Forests Restoration Act to the Collaborative Forest Landscape Restoration Act. All accepted the need to restore fire as an ecological process, and nearly all got their increased burning through wild, not prescribed, fire. What proved popular was mechanical pretreatments, which looked like stealth silviculture. On the Blues the cutting mostly got done, the burning mostly didn't. Fire (and smoke) were not so easily controlled. It was simple to insist that the axe went first—it could prepare sites, it could help pay for other treatments, it fed into the wood products infrastructure. But somehow the burning, which did the hard ecological work, went missing except in piles and in wilderness areas like the Eagle Cap where natural fires were allowed some room and eventually burn scars began to reshape the regime. (There was one novelty for the West: because of WUI issues, some prescribed fire was allowed in part of the wilderness.) Now new management plans are underway to "finish the job" and do the burning. Maybe. Over the next few years the compounding cycles of beetles and budworm and fire are primed to return.[10]

The events in the Blues are unusual in having their saga written in a model environmental history. It's worth pausing to hear what Nancy Langston concludes about the fundamental issues. They have to do with the complexity of the world, our capacity to understand it, and our ability to shape it to our goals. In the Blues every society tried to sculpt the land toward an ideal, and every group for which we have good records stumbled. Foresters could not surrender the belief that, with modern science, they could know, and that knowing, they could bend the woods to their "desired future condition." They "refused to admit even to themselves that they might not know the best ways to manage forests." In a subsequent book, *Where Land and Water Meet*, Langston tracked a very similar story for the Malheur National Wildlife Refuge, south of the Blues, with another cadre of applied scientists, wildlife biologists, taking the place of the Blues' foresters.[11]

The solution, she suggests, is a kind of bold humility, a capacity to act within the face of admitted uncertainty, a vision of life as an endless experiment. Interestingly, this is exactly what Pragmatism, America's contribution to formal philosophy, argued. Ironically, Pragmatism acquired its formal definitions during the same era that created the Forest Service.

======

Irony. For most of the 20th century irony has been the voice of modernism. This isn't the casual irony that is all around us like spores in the air. It's the deep irony that concludes a story or a debate or an understanding. It's irony as a philosophical terminus: it's a necessary irony that means a narrative or an explanation has not truly ended until it comes to rest in an ironic mode. Irony is not something decorative, a vignette like the annotations of an illuminated manuscript. It informs the very structure of understanding. Ignore irony and you will induce sniggers, and maybe condemnation as naive. End with deep irony, however clumsy or misinformed, and you will likely escape criticism. Everyone knows that the world is fundamentally ironic. Virtually all academic history is ironic.

Professional historians know this in the same way foresters knew that wild forests had to be reduced to plantations, and that fire and beetles and budworms had to be eradicated. It's part of their academic training. We can't pretend to know something until its fundamental structure is shaped ironically. That's how we control a narrative. Yet that same urge to control, and belief in our methods to find a way that agrees with our discipline, is exactly what has consistently unhinged remedial actions in the Blues.

In her conclusion Nancy Langston noted that the authors of the *Blue Mountains Forest Health Report* observed how often good intensions led to "potentially catastrophic" outcomes and "seem to feel trapped by the confusion of their history: since they do not understand what went wrong before, anything they do now to correct problems may lead to worse results." Here is the voice of irony measuring the gap between goals and deeds. Rather than doubling down on the science, which has only doubled the ironies, she argues that "we need another set of stories—a vision of wild nature that does not exclude people and cows and logs."[12]

Yet another set of stories might make no more difference than another data set if those stories are still shackled by deep irony. What we need is not new stories but another way of storytelling, a reconception of narrative that is not informed by deep irony. If foresters couldn't look beyond their training, the same might be said for historians. To challenge the models on this level goes beyond substituting one set of prescriptions for another or swapping one story for another. It goes to the identity of what makes one set of practices forestry and others just doing stuff in the woods, or what makes one narrative real history and others just so much anecdote and antiquarianism. If what has happened in the Blues needs history to understand the gap between what was proposed and what has happened, then we need a history that allows us to relate those changes over time in usable ways. Ending in irony is an act of problematizing. We need histories that move on to problem solving.

Deep irony arose to criticize the traumas of putative progress and imperialism, of the economic dislocations of global capitalism, of the jingoist nationalisms that had shaped narratives of the 19th century but in the 20th turned to pathologies. It challenged an older history informed by Providence and Progress. But our needs now are different. No less than foresters, or wildlife biologists, or ecologists in the Blues, historians need to find a working narrative that doesn't simply compromise between the wild and the wrecked but that frames the entire discussion in other terms altogether. There is no privileged perch by which to view the full panorama and not include ourselves—nothing in forestry, nothing in science, nothing in history. That will mean walking away from deep irony. It will be a break from traditional history as profound as New Forestry is from Old.

It's too much to ask for a post-ironic culture, but maybe not for a post-ironic narrative in which paradox replaces irony and working narratives overwrite positivistic and providential ones. We have classic histories from the 19th century that celebrate pioneering and progress. We have histories from the 20th, informed by deep irony, that tell of confusion and decline, of progress in reverse. The 21st century asks for something different. The Blue Mountains might be a good place to begin.

AN ECOLOGICAL AND
SILVICULTURAL TOOL

Harold Weaver

I T'S NOT UNUSUAL FOR someone to be imprinted with the landscape of his childhood and to calibrate fire from that remembered baseline. This was certainly true for the conservative revolutionaries at Tall Timbers, and for many who sought to restore prairie, longleaf, and even pitch pine. But it is odd to the point of quirkiness that that childhood should begin at a camp devoted to hydraulic mining in a side valley of the Blue Mountains. Yet that is what happened to Harold Weaver, who spent his adulthood trying to protect the ponderosa pine forests he had known so intimately in his youth at Sumpter, Oregon.

Harold Weaver is the oft-overlooked member of that great triumvirate that argued in the 1960s for controlled burning. Ed Komarek had Tall Timbers Research Station, and could host fire ecology conferences that ranged the world, and published proceedings to libraries everywhere. Harold Biswell had the University of California-Berkeley, demo sites at Hoberg's Resort and Redwood Mountain, and student acolytes. Harold Weaver worked for the U.S. Indian Service (later, Bureau of Indian Affairs, or BIA) on tribal lands in Washington, Oregon, and Arizona. He was quiet—a forester and an administrator, not a scientist—who published his observations in professional journals, preferring to take photographs, mostly while wandering the woods barefoot. Being in backwater sites, he was able to experiment and observe in ways that would have been unthinkable in the Forest Service; if he did not receive much support, neither did he meet much obstruction. When he left a reservation, its prescribed fire program generally withered. Yet in one critical respect he had leverage that the other prophets of controlled fire didn't.

He was a credentialed forester. He could publish in the *Journal of Forestry*. He could speak, colleague to colleague, on fire and forests in ways that Herbert Stoddard and Ed Komarek, wildlife biologists, or Harold Biswell and S. W. Greene, rangeland scientists, could not. Harold Weaver had professional standing. That didn't mean foresters would agree with him or even listen closely. It did mean that, in a professional sense, they were willing to let him talk.

He grew up in the town of Sumpter, tucked deep in the Blue Mountains west of Baker City. His father, Amos Weaver, was a partner in a cluster of placer claims whose special contribution was to begin hydraulic mining operations "at the earliest possible date in the spring, after winter snow could be cleared from about five miles of water ditch and wooden flumes along steep mountain sides." When the water flowed, it was directed against gold-laden gravel "near the Blue Mountain summit, washing the sludge through sluices." The project continued "day and night" until around July 4 when the snowmelt was exhausted and the flow became inadequate.[1]

Those childhood memories stayed with Weaver. He recalled the "new, bright green needles" of the western larch, which remained one of his favorite trees. He recalled family picnics and camping trips, once the "furious storm of mining" ceased. He remembered gathering mushrooms "in a recent burn" and, later, huckleberries. From about the age of 12, he was "permitted considerable latitude in exploring and roaming with my dog and .22 rifle." Most of the lower elevation forest was mature ponderosa pine, and with help from a narrow-gauge railway, it had been cut over. Still, there were pockets of "mature pine with open, park-like, pinegrass covered forest floors."[2]

He decided to make the woods his career. By the time he graduated from high school, with "interludes" in Indiana and Southern California, those magical forests of the Blues were drawing him back. He enrolled in the forestry program at Oregon State College, spending summers cruising timber in eastern Oregon and California. He graduated in 1928, soon afterward joining the Forestry Department of the U.S. Indian Service, which sent him to the Klamath Indian Reservation in southern Oregon. He stayed with the agency for all of his professional life. In 1933, a year after he married Jessie ("Billie") Gray, he was transferred to the regional office in Spokane to oversee the CCC projects on reservations. From 1940 to 1948 he worked at the Colville Agency, then moved to Phoenix, Arizona, as area forester. Three years later he accepted a position at the Washington, D.C., office as assistant chief, Branch of Forest and Range Management. He returned to the Northwest in 1954 as area forester, stationed at Portland. He retired in 1967, remaining in Portland until Billie passed away in 1978, before moving to Jasper, Arkansas. He died in 1983.

This is not the career of an agency pariah. It's the story of a man who was given serious responsibilities by the BIA, who moved up the bureaucratic food chain, who retired where and when he wanted, with honor. The one anomaly was his passion for prescribed fire. Whatever misgivings the profession and agency had about it, those concerns did not affect his career, nor did they stop him from implementing programs wherever he went. The folklore of the fire revolution sings a saga that tells how its prophets were denounced and scorned by the fire Establishment before they finally succeeded in overcoming their critics. That's not the biography of Harold Weaver.

As a child in the Blues, he had seen fires, which seemed benign compared to the havoc of hydraulic mining and clear cutting, and had foraged among their scars for berries and mushrooms. At Oregon State, however, he was taught ("thoroughly imbued" with) the utter "incompatibility of pine forestry and fire."[3]

The clash between field and classroom worsened on the Klamath reservation, when he met older woodsmen who regarded fire exclusion as foolish and mistaken, though none could answer academic forestry's charge that even light fires prevented regeneration. Light surface burning traded the forest's future for today's convenience. Later, a forester and entomologist, F. Paul Keen, "shocked" him by echoing the same concerns as the "nontechnical" woodsmen, and then took him into the woods to show the antiquity of fire scars amid early-growth ponderosa. That demonstration Weaver then continued on his own "throughout most of the western states," particularly in ponderosa pine regions, which were also where his assignments with the BIA took him. Then he witnessed a wildfire in 1938 on the Warm Springs Reservation that swept through "many thousands of snags and windfalls" in beetle-killed ponderosa. Yet some saplings survived that harsh pruning, flourished, and grew amid a setting of very low fire hazard. That trial by fire convinced Weaver that fire was an ecological process fundamental to ponderosa pine forests and that "fire, under proper control," could be a useful tool.[4]

He published his first report in 1943. It mattered that he framed the issue in terms amenable to forestry. Controlled fire was a silvicultural tool. It reduced fire hazards, promoted growth, pruned thickets—did what foresters tried to do with other methods. It could, in the hands of foresters, assist fire protection and timber production. That didn't mean the profession accepted his notions: most didn't, and the BIA insisted on disclaimers. ("This article represents the author's views only and is not to be regarded in any way as an expression of the attitude of the Indian Service on the subject discussed.") But that year the U.S. Forest Service allowed the Florida National Forest to use controlled fire, along the lines of the "prescribed" burning proposed by Paul Conarro a year before. Still, Weaver's article was a controversial enough notion that the journal solic-ited a formal commentary from Arthur A. Brown, member of the Society of American Foresters and head of fire research for the Forest Service.[5]

Brown conceded that "Mr. Weaver offers some challenges to fire-control policies of public agencies that deserve careful consideration." He worried that Weaver left himself open to criticism by taking in too broad a territory; by speaking to one species when others that, in the absence of fire, could overtake it might prove more economically valuable; by urging a tool, fire, so rife with "ifs" that its effects were hard to predict; by proposing a long-term burning program in place of a temporary boost in fire control that might be enough to produce the desired results. Mostly, Brown concluded, with bureaucratic caution, that "the answer requires considerable prior research."[6]

That is, Mr. Brown sought to reframe the title of Mr. Weaver's article from "ecological" to "silvicultural" factors. "In conclusion," he wrote, "a word on the general philosophy that I believe has controlled to date."

To serve society, the forester must substitute harvesting by logging for nature's method of harvesting by bark beetles and fire. To do that he *must* intervene in the old natural cycle. The first urgent step was to control fire and insects. With nature's harvest reduced, there is an opportunity for system in the second step, which is management of the forest by methods of cutting. With both under full control, which has not yet been attained, there will much room for refinement of method.[7]

Fire control now, control by axe later—a national theme but one that resonates especially well in the Northwest. The argument over fire use was not allowed to spill over out of forestry into realms of ecology, wildlife, rangelands, and so on that the most ardent critics of fire suppression had promoted. Weaver's challenge was treated as an internal quarrel among a band of brothers. Yet without that fraternal badge it is doubtful Harold Weaver would ever have been heard at all.

The first article, "Fire as an Ecological and Silvicultural Factor in the Ponderosa-Pine Region of the Pacific Slope," contained the gist of all that followed. Weaver believed "that periodic fires, in combination frequently with pine beetle attacks, and occasionally with other agencies, formerly operated to control the density, age classes, and composition of the ponderosa-pine stands." Yet 30–40 years of fire exclusion had "brought about changes in ecological conditions which were not fully anticipated, and some of which seem to threaten sound management and protection of ponderosa-pine forests." The proliferation of litter, windfall, and understory thickets had created conditions similar to postlogging slash; Weaver's broadcast burns were promoted to reduce hazard over that dispersed debris. He concluded that "progress in converting the virgin forest to a managed one depends on either replacing fire as a natural silvicultural agent or using it as a silvicultural tool." Weaver found "little evidence of success" in attempts at the first option, and "far too little thought and research" in applying the second. He followed his thesis with examples and evidence.[8]

The examples continued through the next 16 years. In 1947 he argued for "Fire—Nature's Thinning Agent in Ponderosa Pine Stands." In 1955 he considered "Fire as an Enemy, Friend, and Tool in Forest Management." In 1956, taking his case to a wider public in *American Forests*, he noted that "Wild Fires Threaten Ponderosa Pine Forests," and so needed tame surrogates. In 1957, back to the *Journal of Forestry*, he framed the case in more contemporary language, "Effects of Prescribed Burning in Ponderosa Pine." And in 1959 he summarized his career's conclusion with a case study of the Warm Springs Indian Reservation. The ecological damage evident in the pine stands was the result of too much grazing and too little fire. He repeated his fundamental argument: if foresters could not replicate fire in its ecological fullness, they would have to use fire.

His years in Arizona drew particular attention. The program at Fort Apache got front-page, two-photo treatment in the *Arizona Republic* under the headline, "Fire Tested as Forest Friend." The journalist, Ben Avery, had visited an experimental burn in the Bog Creek area and compared it with a wildfire near McNary, "fed by old rotting logs and limbs in a cut-over area." The contrast was convincing. "We always have opposed forest fires. We also have been against sin. But when fire is used as a tool, the evidence of its usefulness was there to see." When the newspaper reported on Weaver's transfer to the Washington office, it noted that "his experiments in utilizing controlled fire as an aid in silviculture . . . have won wide attention."[9]

When the fire revolution arrived, its partisans, eager to find examples of successful burning, honored Weaver as a forgotten prophet and identified Fort Apache as the stellar example of prescribed broadcast burning in the West. But he was not unknown: it was rather that his ideas had not gone beyond his experiments and most did not survive his administrative transfers elsewhere. In 1967, nearly 30 years after the fire on Warm Springs that triggered his curiosity, he was asked to recapitulate his thoughts for the seventh Tall Timbers Fire Ecology Conference. He drew primarily on his experiences in the Colville Indian Reservation in Washington and the Fort Apache Reservation in Arizona. Later, by invitation, he visited Yosemite and Sequoia-Kings Canyon National Parks to comment on their fledgling prescribed fire programs.

In 1972 the Tall Timbers Research Station organized a task force to inquire into the state of burning in the Southwest and invited Harold Weaver to join them. The objective of the tour was twofold. One, it wanted to highlight what could be accomplished in the Southwest's ponderosa pine. ("More controlled or prescribed burning has been done on these three Reservations (mainly the Fort Apache) and over a longer period of time than in any other forested area of the western United States." The primary purpose was hazardous fuel reduction.) Two, it sought to show what could happen if that burning ceased. Weaver had left in 1951; his successor, Harry Kallender, who built out the experiments into a steady program, retired in the late 1960s. The program's momentum kept it alive, until, with a change in tribal leadership, it began to slide into decay; and wildfires returned with a severity not seen before. The 1971 Black River fire and savage Carrizo fire were the precipitating events for the task force.[10]

After those tours, Harold Weaver, well retired in Portland, receded as a personal presence in the fire revolution. He never broadcast those experiments as part of a wider program of fire restoration. He let others add his experiences to the campaign. What endured were his string of publications from 1943 to 1959.

The strongest voices of the fire revolution came from the Southeast. Harold Biswell transferred that chorus to the West Coast, and prairie folk argued for fire in their tallgrass swales. Most of suppression's critics were wildlife biologists, ranchers, wilderness enthusiasts, and park managers. Foresters joined reluctantly then in the Southeast avidly, to support longleaf and loblolly pine plantations. The movement stalled crossing into the arid West, where it stirred memories of light-burning and "Paiute forestry." Harold Weaver mattered because he was both a westerner and a forester

Since the late 19th century, foresters had controlled state-sponsored fire protection. While forestry's European founders did not regard fire as part of their script—fire control was a precondition for forestry, not an ongoing charge—fire protection was what made public forestry powerful in the United States. It gave forestry agencies something visible to do. It gave them a story that could unfold within days, not over generations. It simplified the message of forestry-led conservation in ways the public could understand. Early on, American foresters embraced systematic fire protection as their contribution to global forestry.

They resisted messages from outsiders. In the 1940s and 1950s forestry was agog with industrial logging as the national forests opened up. Foresters doubled down on fire protection as vital to spare that resource for the axe. When they, as a group, converted to environmental sensibilities, they read Aldo Leopold, a forester (Yale, 1909). When, collectively, they contemplated a change in fire policy, they searched for an equivalent figure and lit on Harold Weaver and his string of at-the-time-eccentric articles in the *Journal of Forestry*. They had a prophet from within, and they embraced him.

There is a Pacific Northwest story here, too. In Weaver's work controlled burning didn't challenge the axe: it enabled it. The deeper protest, the sense that fire had an ecological purpose beyond silviculture, had to wait until the second timber wars and the emergence of New Forestry. Fire and axe, green over black—even amidst the fire revolution, the Pacific Northwest held to its originating traits.

EPILOGUE

The Pacific Northwest Between Two Fires

OVER THE COURSE OF the 20th century, the fire scene of the Pacific Northwest seemed to have inverted. The abundant surface burning of eastside forests was largely gone; the explosive fires of westside forests had been tamed into slash piles. The great fires that had ravaged settlements had passed into history. The Tillamook Burn cycle had faded into a green memory. Smoke replaced flame as a public issue. Those who pondered the scene worried more about the lack of good fire than about the threat of bad fire. No one appreciated that the Silver Complex announced a new cycle of big burns, or that America's recolonization of its rural lands awaited the spark of a new era of settlement conflagrations.

It was in this context, and as the second timber war took to its trenches, that three books surveyed the panorama. One viewed fire through the perspective of silviculture, one through ecology, and one through the smoke that was fire's inevitable byproduct.

The first, *The Burning Decision: Regional Perspectives on Slash* (1989), edited by three academic foresters, was the proceedings of a symposium on slash burning. Smoke had become a visible emblem of accelerated logging and, for many urban residents particularly, a public nuisance and health concern. "The decision to burn or not to burn is a major policy issue for modern forest management," argued the editors. But "what constitutes a rational decision?" If smoke was a surrogate for a public debate about the place of industrial logging, then the appeal to a "rational" decision was a surrogate for who would decide. To the authors, "rational" was a synonym for "science," and forest science was embedded in forestry schools. But the issues ranged far beyond those intellectual precincts, which for forestry made them appear irrational.[1]

The second, *Natural and Prescribed Fire in Pacific Northwest Forests* (1990), was edited by foresters, both academic and public, and sought, as had Harold Weaver, to situate burning within a silvicultural matrix. A preface cited Leo Isaac who had described fire and Douglas fir. "Our beautiful virgin forests of Douglas-fir followed fire," Isaac affirmed. With intensive harvesting, controlled fire—slash burning—could emulate that ecology and support reforestation by reducing hazard and preparing sites for planting. Foresters burned postharvest plots as farmers did stubble.[2]

The third, *Fire Ecology of Pacific Northwest Forests* (1993), was written by James Agee, a professor of forest ecology at the University of Washington and coeditor of *Ecosystem Management for Parks and Wilderness*. Its informing concern was not commercial forest but "natural areas of the Pacific Northwest," whose structure also reflected "a past disturbance history that includes fire." The book opens with a discussion of "natural fire regimes." Its core insight was the "seeming paradox" by which "in forest types where fire had been historically infrequent, mandatory fire use became institutionalized (the 'westside story'), while in forest types where fire had frequently underburned the forest, fire use was outlawed (the 'eastside story')." Wildland managers, he argued, needed to restore fire as they might wolves or grizzlies.[3]

The larger setting for fire was shifting. Even as the three books went into print, the second timber war ratcheted toward a climax, and finally an uneasy truce.

———

The 1987 fires were less a long-delayed afterthought than the slow pivot to a new era. A year after Agee's book, the year of the Northwest Forest Plan's record of decision, lightning kindled 99 fires in Washington. One, the Tyee Creek fire, exploded across 135,000 acres and 19 homes, the largest fire on record for the Wenatchee National Forest. The complex became, for locals, Firestorm 1994. In 2002 the Biscuit fire burned 500,000 acres, the largest fire in Oregon since 1868. In 2013 Oregon passed a Wildfire Protection Act to beef up initial attack. Then the 2014 and 2015 fire seasons swept over eastside landscapes.[4]

For the 2014 season the Pacific Northwest sat 43 days atop the National Incident Management Situation Report—a record. For 31 days it ranked at Preparedness Level 5—another record (the previous was 24). Oregon had 58 large fires, Washington 35. Oregon burned a total of 846,945 acres, Washington 413,143. Two complexes claimed the bulk: the Buzzard complex for Oregon at 395,747 acres, the Carlton Complex for Washington at 256,106 acres and 300 homes. Most of the fires started from lightning. As great Aunt Augusta might have put it, to suffer one bad season looks like misfortune; to suffer two looks like carelessness. Big fires were moving from anecdotes to statistics.

Then the 2015 season slammed the region. Washington suffered the largest burned area in its history, over a million acres; at 104,782 acres the Okanagan Complex blew beyond the 1902 Yacolt burn. Three initial-attack firefighters died outside Twisp. Some 630,000 acres burned in Oregon, with two complexes in the southwest, and the main suite sprawled around and across the Blue Mountains; the Canyon Creek complex attracted the most notoriety. But raw numbers don't make fires or fire seasons significant. Their impact with society does.

The collateral damages moved from backcountry to city. Wildfire smoke, it seems, could smother communities as fully as field burning and slash fires. Smoke became an annoyance, and then a public health concern, and it reminded people far removed from the fires that big fires had returned. Then there were economic impacts, not only to timber and forage, but in homes burned, evacuations, and the costs of suppression. Even with Federal Emergency Management Agency (FEMA) grants the charges were high. Washington passed a $178 million supplemental budget to cover state expenses. For the third year in a row, Oregon burned through its normal general fund and Oregon Forest Land Protection Fund, and then its $25 million insurance policy with Lloyd's. The state had long pursued fire protection as a guarantee to those willing to invest in its wood products industry that their lands were secure, and formally insuring the program was a visible marker of that commitment. But three years of drafts threatened to unsettle that unique arrangement, and after 2014 and 2015 the Oregon Department of Forestry had to dispatch a delegation to London to argue the case (which they did, successfully).[5]

Then came the fires of 2017, ending with the Eagle Creek fire started by a teenage boy tossing fireworks off a cliff while companions filmed the scene. With an east wind behind it, and the Columbia Gorge to channel those winds, the fire raced westward across 48,861 acres to the fringe of Portland's suburbs. The popular story of the fire season was first overshadowed (literally) by a total eclipse of the sun, whose arc passed through Oregon and then by the shattering narrative of the fires that blasted Northern and Southern California. (That California should crowd it out of the national narrative seems emblematic.) But the region's fire agencies did not forget. The costs—$454 million for Oregon—threatened to upend their financial model and the premise that fire control was adequate to warrant private capital's investment in a timber industry. The damages threatened to temper enthusiasm for converting low-value rural lands into high-value exurbs. The original Tillamook fire had burned into the backcountry away from Portland. The Eagle Creek fire burned out of the backcountry toward town. The fundamentals had changed.

The modern era of fire protection for the Northwest began in 1910 with the establishment of the Western Forestry and Conservation Association that brought private and public institutions together to address fire. Twenty-three years later the Tillamook Burn, kindled at a logging site, started the cycle of big fires that seemed to define the midcentury. The contemporary era of fire management might date from the 1994 Northwest Forest Plan with its redefinition of forest values; 23 years later the Chetco Bar Complex affirmed a new cycle of big burns, kindled by lightning, lodged mostly in legal wilderness, while the Eagle Creek fire, started at a recreational site, carried the new order of fire to the metropolis. The region's fires had not gone away. They had only undergone a chrysalis and reemerged. Southwest Oregon lit up again in 2018, and the steppes east of the Cascades were burning as regularly as the Great Basin.

The serial big-fire seasons aroused political interest. Senator Maria Cantwell (Democrat) from Washington, and Senator Ron Wyden (Democrat) from Oregon, led inquiries into the country's fire scene and proposed new legislation. The national role the Pacific Northwest had had in the politics of early 20th-century fire with such figures as Senator Charles McNary, it now reclaimed. Such concern was not inevitable. The country had abundant pathologies, all clamoring for attention, and Congress had shown itself unable (and unwilling) to pay for emergency fire operations, bleeding the Forest Service white. Other regions were hammered by fires beyond their experience and capacity, regions with more economic and political clout. Its big fires, however, were shifting the political center of gravity on wildland fire to the Northwest. Big burns were overtaking spotted owls and salmon as the defining theme of its landscape politics.

There was no guarantee that this new wave of burning would persist, but there was equally little reason to think it would not. All the indices behind the big fires pointed their needles in the same direction. There would be milder years and more severe years, but suppression was no longer sufficient by itself, nor, it appeared, was the traditional political coalition behind it. During the contentious debate about field burning and smoke in the Willamette Valley, it was noted that the fissures were not between the two political parties but between ruralites and urbanites. Urban environmentalists tended to want agricultural burning dampened and wildland burning increased. Ruralites wanted more fires in fields and fewer in wildlands. Compromise would be awkward. All parties would need to share a common vantage point.

The Columbia Breaks Fire Interpretive Center sits in Entiat, Washington, along the gorge of the Columbia and in shadow of the 1970 Wenatchee and 1994 Tyee fires, while looking toward the Carlton Complex. Its signature feature is a suite of three decommissioned fire lookouts, now relocated and rehabilitated. Chelan Butte Lookout boldly guards an entry. East Flattop Lookout serves as a central focus for interpretive programs. And Badger Mountain Lookout, like a wooden aerie, looms above the center and its network of trails.[6]

All three structures come from the era between the two waves of big burns that bracket the fire history of the Pacific Northwest. One wave tracks settlement, and it involves mostly westside lands. The other wave, today's, aligns with a postsettlement scene, and its fires are mostly eastside (the exception is the 15-year cycle on the Siskiyou-Rogue). They memorialize a time when the Pacific Northwest, because of its timing in the history of state-sponsored conservation and its significance to the nation's timber industry, commanded special attention in the national fire narrative and devoted considerable resources to halting fire. During the first timber war fire became the point of agreement among quarreling factions, its control the one article of consensus, and it inspired the Northwest's celebrated experiment in cooperative forestry, which became a national norm. Now they are museum pieces at a time when the region again demands national attention and tries to find in fire another common ground.

The Badger Mountain Lookout, perched above the center, is presently closed to visitors until it can be fully renovated. Because it rests on a bluff, one must look to each side from behind it, or to give the scene a metaphorical cast, from the past to a bifurcated future. To the west, you see wildlands; to the east, a strip of urbanization and farmland that lines the Columbia River. Or to put it still differently, there is no single vision possible from the past. Both coexist: neither the wild nor the working landscape dominates. True to its mission, the Columbia Breaks Fire Interpretive Center argues that, in the face of escalating fire threats, everyone should unite behind a strategy to halt dangerous fires, while encouraging beneficial ones. Yet the perspective from the highest site, with the longest view, blocks that unified panorama. Until we can actually enter the structure, the perspective splits into incommensurable scenes.

How to renovate its inherited system to new purposes is the challenge to the fire community of the Pacific Northwest. It has plenty of resources and infrastructure. It is rich in universities and experiment stations and fire lore, and in recent years it has experienced fire's return with a vengeance. The idea that fire could fuse bonds between groups otherwise barely able to speak to each other is not new. It was the core compromise that quieted the first timber war. What made that consensus work, however, was that industry agreed and helped finance the project. The agreement came not out of public spiritedness so much as the threat of a legislated mandate if the quarreling parties did not resolve their differences among themselves.

The hope is that once again the region will rally around fire—its wild presence, its debilitating absence—as a threat to its way of life. Volunteerism, nongovernmental organizations, coalitions of convenience, and foundations offer a shadow semblance to the alliance that had quelled the great fires of settlement. Something analogous will have to happen to calm those of recent years, no longer gorging on the offal of the timber industry but on overgrown, often unhealthy forests, increasingly fenced in by exurbs and wilderness.

What is missing, however, is the active engagement by industry. In the first timber war industry decided to "to find out what is the right thing to do and then go ahead and do it regardless of whose interest it may affect," as George S. Long, manager for the Weyerhaeuser Timber Company famously put it, and threw its influence behind conservation through fire protection. Behind that resolve was the threat of political legislation, but leaders in industry sought a common cause. In the polarized America that emerged during the second timber war, common ground was hard to find, industry and nongovernmental organizations were locked in trench warfare, and the state was effectively neutered. Without the muscle and money of industry, however, the new alliances will struggle to muster the capacity to fashion appropriate fire regimes at the scale needed.[7]

This likely means some compromise must be found that will reconcile fire with axe. In the era of settlement fires, the axe helped power those bad burns. In the era of postsettlement fires, axe and fire have mostly stood apart. The region appears, like some American Nataraja, with an axe in one hand and a torch in the other, while all around flames flicker and soar. The challenge is not just to balance the two, but to make each work with and for the other. It's a challenge the region has faced before and resolved. It needs to do it again.

NOTE ON SOURCES

THE PACIFIC NORTHWEST has a rich tradition of environmental histories, settlement histories, and fire research. Because so much of that history is forest history, which for the region means fire history, I relied mostly on fire texts and institutional studies. The early years I covered in *Fire in America*, and while new scholarship always becomes available, the recent years are very recent, not yet digested. Three fire texts served as background references—Robert Boyd, ed., *Indians, Fire, and the Land in the Pacific Northwest*; James K. Agee, *Fire Ecology of Pacific Northwest Forests* (1993); and John D. Walstad et al., eds., *Natural and Prescribed Fire in Pacific Northwest Forests* (1990). Nancy Langston's *Forest Dreams, Forest Nightmares* (1995) was indispensable for unraveling the history of the Blue Mountains. Gerald W. Williams's synopsis, *The U.S. Forest Service in the Pacific Northwest: A History* (2009), telescoped much that agency's story. I found the Oregon Department of Forestry's website notably rich.

A missed opportunity, one of many because of the decision to treat the region with a minisurvey, is the absence of literary references. The region has its poets and chroniclers of fire. They'll have to wait for another venue.

ALASKA

A FIRE SURVEY

"Alaska is different."
—ANYONE CONNECTED WITH ALASKAN FIRE

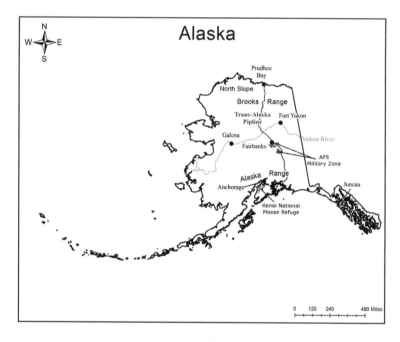

MAP 3 Alaska.

AUTHOR'S NOTE

Alaska

B Y FALL 2015 *To the Last Smoke* had run out of money. So when Ron Dunton approached me at the International Wildfire Conference in Korea about including Alaska in the suite, I agreed if funds were available. Eventually they were, thanks to Ron, the Bureau of Land Management, and the Joint Fire Science Program.

I had a fair collection of historical materials, gathered for *Fire in America*, that carried the Alaska story into the mid-1970s. But the Great Instauration that the Alaska National Interest Lands Conservation Act catalyzed lay outside my previous research. Fortunately, the Alaska Fire Service had commissioned a thorough history of the program, which carried the chronicle into the 21st century. There seemed enough historical materials to stiffen what I hoped to learn from interviews and site visits. The Alaska fire community is relatively small and coherent. I felt I could write a serviceable minisurvey.

By the time the grant came through, I was already committed to a fire tour of the Northeast, and could not visit Alaska until late May 2017. This was pushing against fire season, or at least active preparations for the season, but that was what I could do. I arranged for a two-week visit, during which the Alaska Fire Service (AFS) agreed to send me to some outlying bases. In a good news–bad news story, the weather in western Alaska remained wet and unsettled, so an extended overnight tour morphed into a couple of day trips, but that weather system also held off the fires. The first big fire of the season, at Tok, broke out on my last day in Fairbanks. It made a fine salutation to what fire in Alaska is like, and a nice valedictory to my too-brief trek there.

PROLOGUE

Last Frontier, Lost Frontier

B IG, REMOTE, BOREAL, AND unsettled, maybe conflicted, in its institutional infrastructure—stir these ingredients over a flame, and you get a fire scene unlike any other.

None of these traits is unique to Alaska. What makes Alaska special is how they come together in a mostly fire-prone environment and an often-incendiary politics. Other states have large firescapes, Hawaii is as distant from the contiguous 48, the Lake States have large boreal biotas, and Maine, wholly one. Many states, maybe all of them, have uneasy relationships with the national government, and a handful have cultivated suspicion into a political art form. But only Alaska holds all these traits in a common valence and in exaggerated forms.

Ask anyone in the Alaskan fire community about the state and they will begin with the observation, "Alaska is different."[1]

Alaska isn't just big: it's a veritable subcontinent. It's larger than Texas and California combined. If Alaska were a country, its 663,300 square miles would place it 26th by size, between Columbia and Ethiopia. In John McPhee's memorable phrasing, it's "a place so vast and unpeopled that if anyone could figure out how to steal Italy, Alaska would be the place to hide it."[2]

From Ketchikan to Point Barrow is 1,328 miles by air, as far as from Los Angeles to Chicago. From Ketchikan to Attu is 1,890 miles, or between Los Angeles and Pittsburgh. The North Pacific laps against its southern mountains, the Arctic Ocean against its north slope. Alaskans revel in the number of superlatives the Great Land bestows—the biggest state by more than a factor of two, 17 of the 20 highest peaks in North America, a coastline longer than the rest of the United States combined. Alaska has more glaciers, more volcanoes, more earthquakes; more open space, more bald eagles, more brown bears; and in some years, more burned area.

Its actual firescape is much smaller than the state. Dismiss its two relatively incombustible outliers, the Aleutian Island chain and the temperate rainforest of the Southeast. The remaining landmass divides into east–west trending mountains and plains. The North American cordillera that wends along the Pacific Coast ends as three mountain chains jut westward, one after another, crossing Alaska. The southernmost is the Alaska Range, a vast battlement that arcs into the Aleutians. Above it a stubby range wedges west before dissolving into smaller chains and hills and a vast intermountain interior, a region higher, narrower, and forested in the east, and lower, broader, and marshy to the west. Small ranges, tumbling hills, and river systems, also trending east and west, texture the plain. Farther north, spanning the state west to east, runs the Brooks Range, whose northern flanks spill into the North Slope and the Arctic Ocean.

Alaska's informing firescape is that central region between the two great bounding mountain ranges. Some fires start on the Kenai, and some dapple the tundra above the Brooks Range, but the gravitational power of Alaskan fire comes from the great burns of Alaska's intermountain core, which make the interior into a red giant among the nation's constellation of firescapes. Here summer thunderstorms move inland, following Alaska's great rivers from delta to headwaters. A gradient of dryness and elevation runs west to east. In the lowland west McGrath has an annual rainfall of 18 inches, and 97 of snow. Near the center Tanana has 12 inches of rain, and 44 of snow. In the higher east Fort Yukon village, along the Yukon River, averages 6.57 inches of rainfall a year, less than Phoenix, Arizona, along with 42 of snow. Still, the dimensions are vast, around two-thirds of Alaska, a swath perhaps 600 by 700 miles, a combustible landmass larger than Texas or the Interior West. That's a lot of country with a lot of fires, and they are hard to get to.

───────

Alaska is not just big, it's remote. Fairbanks is as far from San Diego as San Diego is from New York City. It isn't possible to drive to Alaska and stay within the United States; most of the Alcan Highway traverses Canada. Alaska's Little Diomede Island is within sight of Russia's Big Biomede Island. Its historic entrepôts, Seattle and San Francisco, are 1,447 and 2,018 air miles away. But its distance from the nominal metropole is only half the issue.

Its size also makes most parts of Alaska remote from other parts. From Juneau to Barrow is as far as from Orlando to New York. There are few roads (and one railroad) relative to its landmass. Alaska has slightly more mileage of public roads than Vermont, slightly less than New Hampshire. Outside that tiny road network, transport is by barge on the large rivers and by air everywhere. Its capital, Juneau, can be reached only by boat and air. That makes everything expensive, which further adumbrates the remoteness that is Alaska's bane and blessing.

Distance makes Alaska exotic, memorable, and, oddly, forgettable. Within Alaska the scene saturates the horizon. Outside it, out of sight can segue into out of mind. Its remoteness can push it beyond the frame of national awareness. Much as it gets mislaid on the national atlas, so it's easy to misplace Alaska in the national narrative. Even within the wildland fire community, Alaska, in most years, is a sideshow that bursts forth, if it burns big, through June and early July. By July 11, typically, Alaska resources are available to move south.

───────

So vast a land inevitably has lots of biomes. There are rainforests, fog-shrouded islands, forests, tussock grasslands, sprawling wetlands, flood-plains, permafrost plateaus, and mountains. Yet all of the biomes that are burnable are broadly boreal. The wetlands are muskeg and marsh. The interior forests are taiga. The tussocks are tundra. Many soils are peat, underlain or intercalated by permafrost.

Collectively they comprise what the Canadian historian Harold Innis termed a northern economy, a commodities commerce based on fish, fur, timber, and (later) minerals. A third of the state's economy is founded on oil and gas. There is little value-added manufacturing; the in-state economy is mostly services, including tourism. Like Australia, floating on minerals and sheep, Alaska depends on minerals and fish. Over a third of the economy derives from the public sector, especially the federal presence. But its northern economy leaves Alaskans vulnerable to two outside forces over which they have little control—the global commodity market and national politics. Increasingly, that duopoly must expand into a triad to include an unstable climate. In time the biota might evolve beyond the boreal but not in ways that will likely leave Alaskans with more hands-on control over their fate.

So it is, too, with nature's economy of fire. The boreal is a place that defies averages. It's a place defined by variance, for which a norm is valueless for planning. The summers are short and sun drenched, which crowds growing and decaying into a riotous few months, and makes fire ecologically essential to liberate scarce nutrients otherwise warehoused into woody stems and peaty soils. But the fluctuation among fire seasons means that some years burn hugely, and some hardly at all. In 1944, 69 fires started; in 1974, 869. In 1964, 3,430 acres burned; in 2004, 6,590,140. Fire either smolders or soars.

What mediates between outside forces and personal lives are institutions. They absorb the stress, buffer the blows, smooth out the lumpiness between what happens outside society and what is felt inside. Inevitably, they get strained; typically, they can be abused and become unstable themselves when the tensions between what people want and what the boreal environment can provide get too great. In Nordic Europe the land is small enough that society can return the strain to the landscape, which gets cultivated to serve a social model. In Canada agencies like the Canadian Forest Service and provincial departments of natural resources become subject to endless reorganizations. In Alaska it's reflected in chronic complaints about federal oversight.

The paradox is that a first-world economy in a boreal environment requires large-scale institutions, either global businesses or national governments. That applies to the economy of fire as to other matters. Smaller entities cannot, alone, master the episodic fire threats or needs. So while it's hard to speak of "zones" of 40 million acres, or agencies that oversee 60 million acres as small, they can be within the context of a boreal environment. The need to get big propels institutions toward corporatism, consolidations, and confederations.

Big as they are, however, land and fire agencies in Alaska cannot succeed alone; nor can Alaska. In fact, its size worsens the issue because transportation costs become prohibitively high. The state needs to unite with a larger entity. It has to bond to the nation overall.

The foundational Alaskan narrative is the struggle to exercise control over the major conditions of economic and political livelihoods. Alaska— Alaskans—can't do this to the extent they wish but that reality doesn't stop frustration with institutions established to intervene or buffer against those outside forces. Alaska can't control climate directly, or the oil market, or the fish market, so it expresses its unhappiness by blaming those institutions nominally established to modulate the effects. Because ownership of its land has been unsettled so long, the federal government makes a convenient whipping boy.

What metamorphoses garden-variety grievances into political mythology and popular anger is that even when Alaska does own substantial lands, it still can't exercise the level of control or economic vigor its citizens demand. Statehood, in 1959, raised expectations that even Alaska's progressive Act of Statehood couldn't resolve. So, as Stephen Haycox has put it, when "statehood did not generate the prosperity advocates had expected and promised . . . Alaskans fell easily into their habit of blaming the federal government." A rabid "antistatism" came to "characterize Alaska's self-identity to an exaggerated degree."[3]

Its relationship to the country at large is itself a source of both security and instability because it is culture that finally defines how fire is managed and what it means. That Alaska is part of the United States lends its landscapes and firescapes a different character than the boreal forests of Nordic Europe, Canada, or Russia. Nordic Europe is not so remote or so vast that European-style agriculture could not move into it, and where agriculture faltered, silviculture could take its place. The Canadian confederation allowed a political arrangement quite different from the American federation. Canadian provinces control their natural resources; American states control only those lands granted to them upon entering the union. The tension between economy and ecology can arc sharply in the United States, where national lands exist within the boundaries of states, in ways that get diffused in Canada. When big fire years overwhelmed even the most muscular of provinces, and forced them to seek alliances, they did so through a corporation (Canadian Interagency Forest Fire Centre), not a government bureau. The Russian boreal all lies under a central government, though one that has aptly been characterized as a tribute-taking state. Fire suppression, mostly aerial, is powerful but spotty and often arbitrary.

So while many of Alaska's tensions arise from its being within the circumpolar north, being big, remote, and boreal also makes for unique tensions within the United States as well. Alaska was for long a colony, then it stutter-stepped into territorial status, and finally it achieved statehood; and by the time it was admitted it came into an American federation that had learned some hard insights from nearly 200 years of state making and the environmental consequences of unrestricted settlement. Those lessons, codified into the act of statehood, spared Alaska some of the worst excesses endured by other states, but also denied it some of the opportunities that other states had enjoyed (and had not infrequently abused). Its political creation, that is, made semipermanent a political dialectic that has suffused nearly all aspects of Alaskan life, the way color can leak from a pair of socks to tint everything in the wash.

Alaskans have been taught to understand their history as a struggle—a struggle against a majestic but harsh natural landscape, and a struggle against outsiders seeking to control them. The first leads to a heroic story of continued pioneering, broadly conceived. This is interpreted as a matter of mostly individual strength and character. The second story speaks to a sordid political contest against institutions, some of them the expressions of big capitalism, most of them representatives of the federal government, which made Alaskan statehood needlessly problematic. That understanding leverages anecdotal frustration into a mythology.

What both narratives share is a sense of Alaska as the Last Frontier. What that phrase means, however, is itself a source of further contention as groups compete to be the rightful inheritor of the mantle. One vision emphasizes the Wild. Alaska is where America's last best nature endures. Here all Americans can experience and relive the encounter between civilization and wilderness that, in their minds, is the fundamental narrative of America. The other vision emphasizes a Wild West. Alaska is the last chance for frontier development, a place where oversized characters can engage an oversized land and extract wealth and freedom. Nearly all the great controversies have pivoted on the conflict between these two visions. Basically, the Alaskan story is the American story gone far north, where the past can be replayed again but with perhaps better outcomes.

What really makes Alaska special, however, is its Native population. Their presence and voice have destabilized the traditional dialectic into a three-body problem in politics. Wild and Wild West could reenact the drama of American settlement and environmental protection but, unlike the Lower 48, the indigenes could not be silenced, eliminated, or removed into reservations. The narrative arc was not simply the story of a Manichaean struggle between frontier development and preservation, or between state and federal government, but a complex negotiation among state, feds, and Natives. Alaska's Native peoples got land, economic institutions, and a say in how both federal and state lands were managed. This has made the old drama harder to reenact and the old narrative trickier to update. The narrative arc is bent by passing through a prism of Native political power.

———

All these features manifest themselves in landscape fire. How could they not?

Alaska's fire economy is boom or bust, sometimes a sink, sometimes a source. In some years Alaska dominates national statistics, while in others, it has a fire load comparable to that of Massachusetts. Its remoteness means fire control, later fire management, must be conducted by air, which greatly increases costs but also allows for options not often available in the contiguous United States. Aerial fire management of big fires in a big backcountry has sparked a fire culture with its own axioms, initiations, and élan.

A thumbnail history of Alaskan fire has it lagging national developments until 1980, then leading. With statehood it developed within 20 years a sophisticated suppression organization that had taken the nation 70; then it distilled into a handful of years a fire revolution that has struggled, incompletely across the rest of the country, over the past 50. Only Alaska, and maybe Florida, succeeded in restoring fire at a landscape scale. After passage of the Alaska National Interest Lands Conservation Act, Alaska showed what was possible in wildlands.

Its awkward, seemingly never-complete history of land tenure kept Alaska open to changes in fire philosophy and policy. Its relatively segregated realms of anthropogenic and natural fire left it with room to maneuver. It could trade the cost of an aerial fire program for the social and political costs of managing fire in lands, like those in most of the Lower 48, which mingled competing land uses into an ecological omelet. With its immense land base, and relatively few owners, Alaska could dilute the complexity that overwhelmed so many efforts to restore fire elsewhere. It could trade land for decision space.

But Alaska is also a state of mind. The freedoms its geography and history have made possible have also defined a memory, continually rekindled, that has frequently fettered its ability to exercise those freedoms. Alaska might be formally a state, but economically, and hence geopolitically, it remains a colony. Political autonomy quarrels with economic dependence. The upshot, what might be termed the Alaskan Persuasion, is as much a part of its fire scene as black spruce, high tundra, and dry lightning.

THE ALASKAN PERSUASION

ALASKANS AND AMERICANS see Alaska differently. They see the sweeping vistas through different glasses, they read the state's peculiar history from different texts, they interpret the meaning of Alaska through different traditions. The Alaskan perspective, what might be termed the Alaskan Persuasion, began as protest against Alaska's long status as a colony, extended probation as a territory, and halting progress to claim the lands allotted by statehood. Eventually that understanding moved from rhetorical stance to institutionalized mythology to article of faith. Whether its view of the past, and the past's influence on the future, is something others outside Alaska can agree with matters less than how it shapes Alaskans' understanding of the present.[1]

A few states have entered the Union in what is thought as the traditional mechanism laid down by the Northwest Ordinance of 1787. Many, however, have some oddity in their experience that continues to influence their relationship to the country overall and particularly to the federal government. Alaska's experience is distinguished by its long limbo—92 years—from acquired land to statehood.

This peculiarity involves more than remoteness and size. California, both big and remote, was acquired in 1848 through the Treaty of Guadalupe Hidalgo and was admitted with the Compromise of 1850. Texas, big and politically remote, achieved independence from Mexico in 1836, became a nation, and was admitted to the United States in 1845. Maine and West Virginia were calved off from older states, not surrendering the public domain to the national government. Utah had sufficient population to qualify for statehood in 1850 but was prevented until 1896 because of concerns over its Mormon character. Arizona became a territory the same year as California but entered the Union as the 48th state only in 1912, primarily because of concern over its large nonwhite population. That Alaska had an unusual pathway to statehood is not, in fact, unusual.

But it was long, and exceptionally fraught, beginning with its purchase in 1867. Russia was not especially anxious to sell, nor the United States to buy. There was not much to attract settlers beyond furs, maritime mammals, and fish. The task of overseeing Alaska was handed to the Customs Bureau of the U.S. Treasury Department. The army and later the U.S. Geological Survey conducted reconnaissance exploration and mapping for 20 and 30 years, respectively, after acquisition. Once gold was discovered, and Alaska swarmed with prospectors, a stronger presence was demanded. That led to two seemingly competing movements. One granted local control by making Alaska formally a territory, first by authorizing civil government in 1884, and then a territorial legislature in 1912. The other brought Alaska under the doctrine of state-sponsored conservation. The U.S. Forest Service acquired national forests, and a dispute over the leasing of coal lands in the Kenai made arguments over the usage of Alaska's natural resources a national political controversy. By 1912 Alaskans had more control over how the state might develop, but not enough to override national interests. It was as though Alaska were a giant Indian reservation, its lands held in trust by the federal government. At the time Alaska had an official population of 64,356 people, with almost all the nonnatives in towns. (Even postboom, depopulated Nevada had 81,857.) Thus was birthed a political posture that remains to this day.

The pivotal voice belongs to Ernest Gruening. Born in New York City, a graduate of Harvard and Harvard Medical School (in 1912, the year Alaska acquired a territorial legislature), then a professional journalist including editorships with the *Nation* and the *New York Post*, Gruening moved into politics as director of the Division of Territories and Island Possessions for the Department of the Interior (1934–39), member of Alaska International Highway Commission, and finally as territorial governor of Alaska, serving until 1953. He was present during the great buildup of World War II—the Alcan Highway, the construction of military bases, the investment by the federal government in the early Cold War when the USSR glowered across the Bering Strait.[2]

He was the epitome of the educated outsider who came into the country and made it a cause. In 1954 politics and journalism fused as he wrote *The State of Alaska*, a trenchant history-cum-polemic that catalogued what Gruening, in a shaming campaign, presented as Alaska's baleful treatment by a federal government that denied Alaska its ability to develop as a normal part of the United States. Only statehood could remedy a political history that was overweening, abusive, and clumsy. He made statehood his mission. In 1955 he was elected to the U.S. Senate as an Alaskan advocate. He enjoyed full senatorial privileges when statehood arrived on January 3, 1959.

Gruening served in the Senate until 1969, and his voice seemed to speak for all matters Alaskan and mapped the discourse that informed an Alaskan Persuasion for what resembled a state-nation like Texas and California, but mostly resembled a state-colony. Its twin geodetic markers, from which all else was triangulated, were land and politics.

———

Hostility to a central authority is not unique to Alaska. It's foundational to the American experiment. Since 1980 it has been a platform for one of the country's two political parties (paradoxically, the one now in power). Different regions, states, and lobbyists offer their own reasons for skepticism and resistance. The South wants to stall civil rights. Wall Street wants to hamper regulators. The oil and gas industry wants to unlock potential reservoirs and curtail pollution controls. Antagonism to the federal government is axiomatic in Nevada, Utah, and Idaho (which have abundant public lands under federal administration) and Texas (which has few). The congregation and the choir are the same.

What makes Alaska's situation different is that it has leverage through the 1958 Statehood Act, which granted Alaska the chance to select 104.5 million acres as its own, and the 1980 Alaska National Interest Lands Conservation Act, which confirmed the process. Other states clamor for federal lands to be ceded to the states (which will likely privatize them). Alaska has had lands coming to it, lands that it could choose. There is something tangible at stake. What has kept the pot simmering is that the transfer of lands has been slow and encumbered by what Alaskans regard as meddling.

Since Gruening it has been axiomatic that statehood would yield lands, and those lands, once developed, would fill the state's coffers and fund Alaska's future. State lands would make the state economically autonomous, or at least allow it to provide for basic governmental services. To ensure that the state would not sell off its lands, and so impoverish itself over time, the Statehood Act, uniquely among states, "forbade the new state to sell or otherwise dispose of the mineral land given to it." Like much of the Alaska Statehood Act, this provision reflected the lessons of history. States without at least a tithe of federal lands tended to be politically subjugated to the economics of commodity production—timber in Maine, coal in West Virginia, oil and gas in Texas (after cotton and cattle). Too little federal land can be as upsetting as too much.[3]

Still, Alaska had the opportunity to select lands of high economic value, an option not available to the previous 48 states. The anticipated scenario began to unfold when immense oil reserves were found on state lands around Prudhoe Bay on the North Slope. To get that oil to market, however, required a pipeline, a construction project as daring in its way as the Panama Canal. Then the political pipeline clogged.

Alaskan Natives demanded equal rights and lands of their own, and a new era of environmentalism swelled to wash over Alaska's backcountry. In 1968 Secretary of the Interior Stewart Udall agreed with the rightness of the Natives' cause and shut down land transfer, which also stalled the pipeline, until the politics was resolved. That came with the 1971 Alaska Native Claims Settlement Act (ANCSA), which granted lands and organized the Native peoples into 13 corporations. The Trans-Alaska Pipeline Authorization Act was signed in 1973. By now, however, a new environmental movement was gathering strength that asserted national claims for conservation. This time the emphasis was not on national forests but on parks, wildlife refuges, and especially wilderness. Since such categories would determine land use, the land cession process again stalled. In 1978 Alaska senator Mike Gravel threatened to filibuster to prevent passage of the reclassification; President Jimmy Carter responded by transferring the at-risk lands into national monuments by presidential proclamation. Resolution came with the 1980 Alaskan National Interest Lands Conservation Act (ANILCA). The State of Alaska ended up with 104.5 million acres, Alaska Natives with 44 million, and the federal government with 225 million acres, two-thirds in designated conservation units and half in legal wilderness.

The conflict—so long it seems almost constitutional—has been defined several ways. Gruening framed it as a question of sovereignty, a struggle to gain control over the land. That remains a sturdy subtext for the prevailing frame, a controversy over how that land should be conceived, between two poles, both of which frame their advocacy around the notion of Alaska as a Last Frontier and themselves as the rightful heirs to that tradition. Preservationists see an unspoiled Wild; developers, a Wild West. Once again, such notions are not unique to Alaska. What makes Alaska's narrative different is that it was destabilized by ANCSA in which Natives were guaranteed rights to a subsistence economy that was neither a capitalist market nor altogether removed from usage. Land development was thus different in the Alaskan context. So was wilderness. And so, as the smoke cleared, was fire management.

Alaskan literature, too, harks back to a Last Frontier. A written literature begins with the Gold Rush. Since then, most texts have been nonfiction, with classics such as Ernest Gruening's *The State of Alaska* (1954) in the buildup to statehood, and John McPhee's *Coming into the Country* (1977), presenting the debate between development and preservation on the eve of ANILCA. Note that both deal with politics and land.

Popular literature is less politically attuned. Its genetic essence is a material contest—struggling to survive, striving to assert one's self in the face of an overwhelming, or at least cantankerous, nature, pioneering toward a future. It overflows with endless accounts, often locally published, of individuals, and sometimes couples, reenacting the frontier saga, building cabins, living off the land, struggling against the cold, the arduous, the wild. Mostly, this is a literature of exceptionalism that celebrates virtues, themes, and story lines no longer possible to reenact in the Lower 48. Jon Krakauer's *Into the Wild* (1997) updated that tradition for high culture with his story of a young man choosing to live off (and ultimately die in) the Alaskan bush. Not many Alaskans elect to live this way, but they celebrate the option.

In his famous public letter on wilderness, Wallace Stegner argued that we need that untrammeled land to remind us, even if never do more than stand at its brink and look. Ardent Alaskans seem to need the prospect of untrammeled agency, whether or not they ever exercise it. In political terms it is not that the federal government is always the problem (its subsidies keep the state's economy from collapsing) but that it challenges the narrative.

Not surprisingly, this is the motif captured in Alaskan literature about fire. While that literature genuflects to place and scenery, it is not specifically rooted anywhere in the land, only to sites with fires, which is to say, where struggle is possible. Rather, like fires, it moves around the larger landscape, even migrating to the Lower 48 if that is where the action is. It considers fire only as a natural nemesis—the flaming equivalent to a blizzard, a metal-crushing cold snap, or a rogue bear; an opportunity to pit one's self against a primal antagonist. There is no discussion about fire's ecology or the way suppressing fires that have sculpted the taiga for millennia might share the narrative of stripping soils for gold and fish-trapping salmon to the point of exhaustion. Fire exists as a challenge, like a deep frost or a too-curious brown bear.

Here is Murry Taylor, giving that sense a comic moment in *Jumping Fire*, the one text to achieve some national attention (italics in original):

> *Shit,* I thought. *The hell with this. Way out here in the middle of nowhere, beat to death, tired, wild-pig dirty, nothing to look forward to except ash pits and smoke, getting your face scratched, maybe your eyes put out. Eating burnt flat nose, drinking bad coffee, shitting in the woods, wiping your ass with sticks and dry moss, watching over your shoulder for bears, and having to stay up all night by yourself with no one to talk to. What a weird fucking way to live. And for what? Another long day tomorrow and the same old shit all over again.*[4]

Weird, perhaps, but recognizably within an Alaskan tradition. It is a job done right, a challenge met, the wild wrestled to Earth. The jumper survives to continue through the season, and into another. That smoke-jumping is a job paid by the federal government to manage public lands—a topic (and literary device) that could drill into the core of Alaskan politics—is never discussed. What matters is the individual struggle to overcome, preferably against some primordial element, and through it, master himself, however misunderstood that instinct might be outside Alaska.

Oil has changed Alaska, and Alaskans. While they cherish a subsistence lifestyle as a literary trope, as they might the sight of Denali, fewer Alaskans each year live by one. Rural Alaska has been in decline for decades, and where villages exist most no longer follow the seasonal and migratory rhythms of the past.

Today the great majority of Alaskans live in cities and suburbs, where they live like urban Americans everywhere. They drive the same cars. They eat at the same fast food franchises. They watch the same TV shows and follow the same sports teams. They shop at malls similar to those in Sacramento and Denver. Commuters suffer through rush hour. Teenagers are tethered to phones. Overwhelmingly they work in service industries. Many are public servants.

They expect the same standard of living they believe characterizes the rest of the country. As Stephen Haycox, dean of Alaska's historians put it, a "more realistic history of Alaska" acknowledges a "population without a self-generated economy that demanded all the services and amenities of contemporary American culture." Put bluntly, Alaskans want a first-world life with a third-world economy. Yet outside Nordic Europe, boreal environments are commodity economies. Alaska has no agriculture, no pastoralism, and no manufacturing. Its economy is closer to West Virginia than to Texas, which does have extensive agriculture, a large population, a high-tech industry, and manufacturing.

Alaskan statehood brought costs as well as assets. As the saying goes, it's better to be lucky than good. Australia, famous as the Lucky Country, succeeded as a modern economy despite distance and a harsh bush thanks to exports of wool and newly valued minerals, in a "rush that never ended." Alaska is lucky only when gold (in the past) and oil (recently) are flush. It becomes angry when the rush goes bust. (The obsession with drilling the Arctic National Wildlife Refuge is more symbolic than practical: the collapse in petroleum prices is the result of fracking, not reserves "locked up" by the feds.) Otherwise the state depends on subsidies of various kinds from the federal government, which makes the feds an easy scapegoat.[5]

On June 1, 2017, I was listening to KUAC, the National Public Radio station in Fairbanks, while it was discussing how the state's fiscal crisis would affect higher education. The president of the University of Alaska-Fairbanks explained that the situation would not be so critical if the federal government had bestowed on Alaska the land-grant awards it had made available to the Lower 48. Always land. Always the federal government.

PYROPOLITICS, ALASKA STYLE

FIRES ARE A GIVEN in Alaska, at times as seemingly abundant as black flies and bears, and have been since the Pleistocene. If Alaska's people disappeared, its fires would still flourish. But because humans—Earth's keystone species for fire—are present, fire regimes are not only shaped by drought, winds, terrain, and forests, but by culture, institutions, ideas, and personalities. Alaska's fires are not just a natural phenomenon but a political one.[1]

That wildland fire protection might be a political project should surprise no one; anything that affects public assets and public safety *should* be political. What Alaska has contributed to the American story is clarity in how fire management relates to land management. In the Lower 48 the story is one of mixed practices and complicated ownerships, knotted together by mutual aid agreements, and an institutional matrix that, until the fire revolution of the 1960s, was overseen by the U.S. Forest Service. In Alaska there were fewer owners, the role of the Forest Service shrank rather than swelled over time, and a wildland fire establishment evolved in syncopation with the peculiar politics of land tenure.

═══════════

The cadences of burning beat their ancient rhythms until a succession of gold rushes upended the old order by adding starts, kindling a decline in indigenes, and scrambling social controls. The strikes began around Juneau and Kenai in the 1880s; they exploded across the interior with the Klondike stampede in 1898; they sprawled into the Seward Peninsula and Nome in the early 20th century. By now other minerals, notably coal, came into prominence. The gold brought Americans into Alaska, but when the gold panned out, the people left. Coal (and later oil) made deeper penetration possible and something like permanent settlement with commercial links to the outside world. A stampede of fires accompanied the rush of newcomers. An observer of the Copper River region remarked that "during the late gold fever flames were to be seen in the summer months on all the mountain sides, where they looked at night like the outpost lamps of a great city."[2]

The gold rush changed the political order. It demanded an administrative presence in the interior; it was not enough to patrol coastal water with Treasury cutters; it forced Congress to reconsider Alaska's political standing. The reorganization occurred amid enthusiasms for state-sponsored conservation. The U.S. Geological Survey was mobilized, and forest reserves were proclaimed. In 1905 the U.S. Forest Service acquired national forests from the General Land Office in the Department of the Interior; that included the Alexander Archipelago Forest Reserve in the southeast. In 1907 the Tongass National Forest was gazetted; a year later it merged with the Alexander Archipelago. The Chugach National Forest was created in 1907. Neither had serious fire problems, though fire control was something the Forest Service did as a badge of its administrative identity. The reserves in the sodden south brought the agency to Alaska, but its perspective on fire would not remain south of the Alaska Range. In 1909 it dispatched R. S. Kellogg to inspect the interior as a prospective arena for further national forests.[3]

As with other matters Alaskan, the political process stalled. Alaska acquired a territorial legislature in 1912; the Treasury Department yielded to the Department of the Interior. There were no further national forests, no moves to organize its internal hinterlands, no efforts at fire protection. Alaska's economy focused on mining, mostly past its flush times, and salmon, approaching a crisis of overfishing. The conservation movement was riveted over the brazen fight between Chief Forester Gifford Pinchot and Secretary of the Interior Richard Ballinger over the leasing of

Alaskan coal lands. So while there was plenty of fire in its intermountain core, there was no systematic effort to understand it or mechanism to manage it or even a sense of urgency that such fires warranted much attention.

So long as settlement remained on the coasts, Alaskan fires remained on the margins. What changed the calculus was a political act, granting the colony status as a legal territory, and an economic one, the opening of the interior by railroads. The two of course were closely related, and both were decisions made in Washington, not Juneau.

The Alaska Railroad began construction in 1915. One terminus lay in Seward, the other, in Fairbanks, with Anchorage as the construction entrepôt. The project moved Anchorage, a flea-bitten hamlet of a few hundred people, into the economic engine and leading city of Alaska. At Fairbanks steam on rails could meet steam on the river. Both opened the intermountain belt to entry with a first flush of fires—fires kindled directly by locomotives, fires encouraged by slash and careless campfires. This is what railway construction did everywhere. Alaska's innovation was that it happened in landscapes prone, in the right years, to explosive burning and amid a settlement so sparse that fire control was a delusion.

The Forest Service took an interest. It dispatched several reconnaissances, including one by Chief Forester Henry Graves, and published appeals, and then a plan, for fire protection that would extend into the interior. The Alaska Railroad had become a corridor of burned land. In 1921 the General Land Office's Division of Field Investigations began patrolling for fire selectively along the tracks and around some towns. The Forest Service pressed for more action, and in 1928 became chair of a Forest Protection Board, a national oversight group to coordinate fire (and other issues) on all federal forested lands. A year later the General Land Office (GLO) patrolled the railway from Seward to Fairbanks, and persisted through 1933. The GLO continued its patrols through 1933. The Forest Service continued to campaign for effective protection.[4]

Then Harold Ickes, secretary of the interior, visited Alaska in 1938 and was appalled at its fires. The next year, at his urging, Congress established the Alaska Fire Control Service (AFCS) and charged it with fire control statewide, with forestry and recreation programs, and, in 1940, with the administration of the CCC program. A field handbook prepared in 1940 boldly proclaimed that the purpose of the AFCS was "to protect the forests and vegetated areas of the Alaskan public domain lands from the depredations and ravages of uncontrolled fires. It is pledged to use its every resource, and those at its command, toward the prevention of fire, and, within its scope of authority and means of practical accomplishment, toward the suppression of all fires regardless of origin." The handbook was written by foresters.[5]

The official fire plan was somewhat more cautious. "The policy includes protection of the forests, woodlands and tundra regions of Alaska and the detection and suppression of fires occurring in and adjacent to the more populated districts." The General Land Office did not yet have an emergency fire account, and "in lieu of limited suppression funds, fires occurring in remote regions will not be touched. Strict adherence to the policy of not acting on a fire unless it can be corralled and put out will be maintained." The AFCS would emphasize patrol along rivers and roadways and the protection of villages. Not until the late 1950s did the BLM attempt blanket coverage over the interior. Meanwhile, what it lost with the termination of the CCC, the AFCS gained by affiliation with the military. With World War II and the subsequent Cold War, Alaska acquired strategic importance.[6]

The interior was again opened up, this time by military occupation. The Alaska Highway was constructed; new bases and airfields were laid out; and air travel to the continental United States was established. The enlarged military presence was especially favorable to promote wildfires (because of hasty landclearing) and to fire control (because of military assistance with men and equipment, including aircraft). The AFCS entered into a wide range of cooperative agreements among state and federal agencies, along with the military. The Cold War strengthened the military presence in Alaska, and cooperative agreements between the military and the BLM assisted the transition to civilian fire protection. The two evolved reinforcing strategies of protection through airpower.[7]

In 1946 the AFCS was absorbed into the newly formed Bureau of Land Management. The next year it was disbanded as a semiautonomous agency and incorporated within the BLM's Division of Forestry. At first it seemed that fire control would be lost in a labyrinthine bureaucracy dedicated to other concerns. In 1952 Starker Leopold and Fraser Darling, who inspected the causes of wildlife depopulation in the interior, agreed, concluding that in central Alaska "it appears to us that range destruction by fire is principally responsible." They noted the "ambitious but sadly underfinanced fire control program" of the BLM and applauded it as the "first positive step to curb this destruction."[8]

There were few dissenters: progressive thinkers and scientists all supported aggressive fire protection as a foundational policy. The early 1950s were big fire years. Slowly the BLM developed a protection force based on civilian aircraft, although the army and the air force contributed heavily at times under cooperative agreements. In 1954 the BLM acquired three Grumman Goose model aircraft from the coast guard and several small planes for detection. In 1955 the BLM released a "Comprehensive Forestry Program for Interior Alaska," in which, naturally, fire protection loomed large. All that was required for a major BLM investment in fire control was a catalyst.

By any standard the fires of 1957 were enormous. Most were lightning caused. At least five million acres burned, but the full extent remains unknown. Suppression forces were quickly overwhelmed. BLM overhead from the Lower 48 were shipped north with minimal success. Smoke saturated the Alaskan skies, shutting down air traffic in the interior, isolating villages, alarming the military, and for two weeks shutting down the airfields at Anchorage and Fairbanks. Forestry magazines publicized the fires as a matter of shameful neglect. The military worried that such conflagrations might compromise the strategic value of its Alaska outposts. Politicians recognized that the territory, soon to be a state, demanded more attention.[9]

Old-timers of the AFCS regarded the 1947 fires as the worst. But those burns hardly sounded outside an Alaskan echo chamber. The 1957 fires rang brazenly through the national media. What separates a big fire from a significant one is how it engages with the larger culture. In 1947 the AFCS was being absorbed into a newly constituted BLM. In 1957 Alaska was on the cusp of being admitted to the Union. A swarm of conflagrations is not how new states clamored for admission, or how the rest of the United States welcomed new members.[10]

Statehood on January 3, 1959, marked a phase change for wildland fire. Within 20 years Alaskan fire would recapitulate what had taken the Lower 48 nearly 70.

Its most visible expression was a breakneck escalation in technology, research, and force available for fire control. The American Forestry Association released "A Fire Plan for Alaska." The Forest Service sponsored research, from reconnaissances to fire-danger rating systems to statistical summaries, culminating in an Institute for Northern Forestry at Fairbanks. Smokejumpers from Missoula did a tour in 1959. Crews of emergency fire fighters composed of Eskimos and Athabascan Natives were organized, with seasonal fire employment a critical source of cash. And aircraft became the technology of choice. Between cooperative agreements with the military and war surplus aircraft, an unrivaled aerial fire program leaped into being almost overnight.

The fear that Alaskan fire would be lost in the BLM was inverted. By 1963 its Alaska fire program was larger than anything the BLM had in the Lower 48. After the 1964 Elko fires, the agency imported Alaskans, notably chief fire officer Roger Robinson, to replicate in the Great Basin what they had done in intermountain Alaska. The project reached a bureaucratic climax when the Boise Interagency Fire Center went operational in 1969, under a BLM director, a public emblem to the agency that it had achieved equality with its old rival the Forest Service. That summer the Boise Interagency Fire Center supported the Swanson River fire on the Kenai, which ran up the largest suppression bill in history to that time.

With two poles to arc between, the BLM continued its breathtaking climb to the summit. It developed a lightning-detection network to record lightning strikes and so direct aerial reconnaissances. It boosted remote automated weather stations. It experimented with new parachutes, with methods of paracargo dropping, with specialty fire engines like the Dragon Wagon. It tested cloud seeding as a means of weather modification to quell lightning and promote rain. Technology could complement aircraft to overcome the distances that made fire protection difficult, expensive, and often ineffective. Especially in Alaska, a better organization could shrink the number of ignitions that became big, but it could never abolish the conflagrations that could wipe out all the years of small fires and lands nominally protected. Worse, the longer the land remained unburned the more likely it would burn big.

That paradox was at the core of national policy reforms, what might be aptly characterized as a fire revolution that sought to replace a singular policy of fire suppression with a mixed policy that would promote good fires while still preventing bad ones. The opening fanfares came in 1962; by 1968, the National Park Service had renounced the 10 a.m. policy for one of fire restoration; in 1978 the Forest Service instigated a full-spectrum program of reforms. The BLM, first in Alaska, then in the Great Basin, ran cross-grained to this chronicle. It still strove to emulate the Forest Service, but its model was the U.S. Forest Service (USFS) of the 1950s, not that of the 1970s. It had, it seemed, substituted technology not only for distance but for policy. It pursued a high-tech, old-school program in fire suppression, willing to draft as necessary from the Lower 48. The Swanson River fire was exactly the kind of extravagant suppression blow-out that critics loved to hate.[11]

More lay behind the revolution than just an ironic buildup of fuels. The public was enthused about wilderness. Wilderness partisans saw in free-burning fire a symbol of a wild restored, wildlife advocates saw in those fires a means to renew habitat. Forestry became a discredited profession, unable to engage with its sustaining culture in much beyond board feet, and forestry's political expressions, most visibly the U.S. Forest Service, slid into decline. This mattered because foresters had long been the voice of fire protection. What the BLM was doing in Alaska reflected norms prevalent during the era of the AFCS and the early decade of state-hood. While the Alaska fire organization could leap to the forefront of technology and firefighting tactics, it was lagging in policy and strategy. Most of what Alaskans said they wanted from their bush would be better served by restoring fire than by suppressing it. The damages left by sup-pression bulldozers and trenched firelines in permafrost would linger for decades.

The deeper current, as always in Alaska, was the question of land ownership. The Statehood Act granted 104.5 million acres to Alaska and 25 years in which to make those selections. The BLM would no longer reign supreme north of the Alaska Range. The state would have a consid-erable estate of its own, and was eager to assume full control over it. The discovery of oil on state lands around Prudhoe Bay promised the wealth to fund those ambitions. But the oil came years after statehood, and money for state programs would lag further, so as a temporary measure the BLM signed a "reimbursable cooperative agreement" with the state to provide fire protection on state lands until the state had the capacity to do it for itself.[12]

State land selection stalled over Native claims. Neither the BLM nor the State of Alaska had negotiated with Alaska's indigenes, nor had the United States signed treaties, and the Native peoples demanded rights to choose lands for themselves. In 1968 Secretary of the Interior Stewart Udall agreed and closed the process of land transfer until those claims were resolved. That stopped the state's acquisitions, and it stopped construction of the Trans-Alaska Pipeline, but it did not stop fire protection. The fires still came, and until the state could establish jurisdiction, the BLM continued its boisterous buildup. The state contracted for fire services ("as an interim measure") on the lands it had so far picked. After Swanson River the BLM received special spending authority to bolster initial attack forces, specifically through helitack.

The 1971 Alaska Native Claims Settlement Act freed the state to restart the transfer of lands and granted Native corporations 44 million acres of their own. They first promised to siphon off suppression from the BLM, but a small clause in ANCSA kept the BLM as the responsible fire agency until the Native corporations could achieve "substantial revenue" sufficient to assume fire protection for themselves. This proved a slippery clause that has left the BLM in more or less permanent trusteeship. These lands were in addition to those selected as far back as the 1906 Alaska Native Allotment Act, which allowed Native peoples to patent as much as 160 acres, which eventually involved thousands of claims scattered across Alaska. The Bureau of Indian Affairs was technically responsible, but north of the Alaska Range it had no capacity for fire protection, so this task, too, had devolved to the BLM.

With the passage of ANCSA, the BLM was directed to "disengage" from state lands around Anchorage, the Matanuska and Susitna Valleys, and the Kenai Peninsula. That led to discussions with Alaska Department of Forestry and a target date of 1975 to end its fire protectorate. In 1973 Governor William Egan directed the Alaska Department of Natural Resources to begin a genuine program in fire protection. A bond issue the next year brought monies to establish facilities, mostly along roads and around towns. Two years later the BLM received a long-missing organic act (the Federal Land Policy and Management Act) and its Alaska bureau signed cooperative agreements with the Alaska Department of Forestry.

Year by year the state assumed operational control over more of its legal estate. It helped that BLM fire officers in state-selected areas often signed on with the Alaska Department of Forestry (ADOF), which assured a degree of expertise and continuity. Alaska's lands, however, were chosen for their urban value and natural resources (particularly minerals), not for ease of protection. The most threatened lands were concentrated along the emerging rail and road system where settlement clustered; these lay mostly south of Fairbanks. The state had the highest values at risk (Fairbanks' fire losses were "the worst in the industrialized world"); the BLM had the capacity to fight fire everywhere. It all made for an awkward alliance, complicated by different styles of firefighting; the BLM was a wildland agency, Alaska Department of Forestry mostly concerned with structures. The state inquired into models of "dual-capability" fire services, of which California seemed the most effective, though also the most expensive. Legal transfer of land was one thing, actual management was another.[13]

Then another land controversy intervened. Section 17.d(2) of ANCSA allowed the federal government to designate "national interest" lands. This set in motion a campaign to reserve much of the remainder of federal lands as legal preserves. The crusade intended to move as much BLM land as possible into national parks, wildlife refuges, and wilderness. This was a national debate about the future of Alaska's still federal lands. The BLM would likely shrink further; at some point it might no longer have the capacity to conduct the kind of aerial fire protection it had spent the past two decades perfecting and then exporting to the Lower 48.[14]

Meanwhile the fires continued. The 1971 and 1972 seasons hit a million acres each, and 1977 blew up into 2.5 million. Big fires, as always, prompted responses. The 1977 season was especially significant because California endured a rolling thunder of damaging fires and the Wenatchee fires in Washington returned big burns to the Northwest. It was, arguably, the last season officially under the old 10 a.m. regime. The Forest Service had to admit, at least to itself, that it could no longer hope to go it alone and that fire exclusion in most wildlands was a self-defeating mistake. Interagency cooperation on large fires, with the Boise Interagency Fire Center (later renamed the National Interagency Fire Center) as a coordination center, was the future of fire suppression. But it was also true that the future no longer looked solely to suppression. A year later, fire restoration went national when the Forest Service overhauled its policies and programs. The reforms were a revolution from above that would struggle to get rooted in the field. Those notions affected Alaska, but inevitably they would be refracted through a looking glass of land ownership.

The outlines of the new order crystallized in 1978. The Forest Service renounced the 10 a.m. policy in favor of a pluralistic approach toward fire that would balance fire protection with fire restoration. The State of Alaska established a fire organization for south-central Alaska, its primary focus. And the BLM formally transferred 204E lands to other departmental agencies. In the Alaskan fire scene the number of players had increased, the chips they played with—their lands—had changed hands, and the rules of the game had been redefined. The only constant was the magnificent infrastructure and trained organization its fervent engagement with fire had bequeathed the BLM. It continued to provide fire services for all of Interior's agencies.

The next year offered further refinements. Alaska DOF and the BLM agreed to a geographic partition of their Alaskan fire charges. The Alaska Department of Forestry would assume responsibility for land in the Fairbanks-Delta area, except military bases, and for lands south of the Alaska Range, except the Kenai Moose Refuge. The Bureau of Land Management would protect the rest. This left ADOF responsible for 19 million acres of state lands and 10 million of federal, and the BLM responsible for 35.6 million acres of state land. Presuppression costs for each party were about equal, but because most of its fires were off-road the BLM had to resort to high-cost aircraft, which left its suppression costs higher.[15]

The feds were also looking both to segregate ownership and to integrate operations. A letter of agreement among Interior's assistant secretaries charged the BLM, National Park Service, Fish and Wildlife Service, and Bureau of Indian Affairs to begin wildfire protection planning for Alaska. Those plans also included lands under Native corporations. On February 13, 1979, a memo from the BLM assistant director assigned BLM Alaska to undertake the necessary planning.

What fire policy might be appropriate had to wait, inevitably, for a resolution of land allotments among the feds. That only came in 1980 with the Alaska National Interest Lands Conservation Act. In many respects ANILCA was a high-water mark for environmental preservation. The act gave legal protection to some 104.5 million acres (28 percent of Alaska, the same sized landed estate the state received). With 56 million acres the National Wilderness Preservation System tripled. The landed estate of the National Park Service doubled.

Those lands came out of the ever-dwindling dominion of the BLM. The BLM, it seemed, was reincarnating the role of its predecessor, the General Land Office, as a temporary custodian and ultimate disposer of the public domain. With the state, it transferred land to another public entity, which in turn would privatize plots or lease rights. Within the federal domain, it relinquished lands to other agencies, each of which had its own mission and would want to create operational capacity to satisfy its goals. A natural expectation was that both the Park Service and the Fish and Wildlife Service would in Alaska, as in the Lower 48, establish their own fire staff in order to promote more free-burning fires.

But that was implausible. It would multiply what needed to be merged. The federal agencies could not, each one, provide fire management over their lands. The crescendo of 1980 was the ascension of the Reagan administration, which sought to shrink the federal government, transfer funds from civilian agencies to the military, and roll back environmental reforms. There would be less for everyone. The coming charges against the feds were more likely to refer to federal underreach than overreach. Nor could ADOF succeed if the BLM fire program imploded. Interior needed an internal arrangement similar to that between the BLM and ADOF. And Alaska needed to consolidate even as it subdivided. The entire system of Alaskan fire had to reboot.

―――――――

The reboot took two complementary tracks. One clarified policy, what was to be done and why. The other track addressed who should enact that policy. Together, almost overnight, they moved wildland fire in Alaska from being an aggressive throwback to the avant-garde of America's great cultural revolution on fire. The peculiar pyropolitics of Alaska created something new under the midnight sun.

A cascade of memos chronicles the developments. On January 26, 1981, a letter to the Alaska Fire Subcommittee from the Alaska state forester suggested a five-year timetable for transferring responsibilities from the BLM to ADOF. On March 30, 1981, the BLM director, BIFC wrote a memo to the BLM director, Washington Office, to "fully support" the proposal, while noting some points of abrasion. The arrangement would demand flexibility since novelty and circumstance would surely take it in directions not anticipated. It would require that the BLM accept an entity, the AFS, that could not "be squeezed into the standard BLM District organization," where it was nominally housed. It would require constant maintenance and commitment among the partners. The Park Service was not interested in continuing the old suppression program: it wanted to reinstate fire in something like its ancient scope and would create its own program if the reconstituted AFS did not do what the agency's mission demanded. The Fish and Wildlife Service presented a "very dangerous situation" in that they "do not believe that they have any fire problems." In fact, they have "both huge problems and opportunities, and the BLM must not get trapped by their self imposed blindness." (The agency's regional fire problems were about to go national after a fatality fire at Merritt Island NWR in Florida; this led to a national fire office that made negotiations easier for cooperating agencies.) Both the National Park Service and the Fish and Wildlife Service implied they had little need for a legacy fire suppression organization. In such circumstances "fluidity" was more than good: it was essential. All in all, the memo concluded that the proposal "will work, and is a good—if not the only valid[—]approach we can make."[16]

On October 16, 1981, a memo from the Department of the Interior to the BLM Alaska state director approved "a single fire suppression and fire logistical support organization in Fairbanks," thus birthing the Alaska Fire Service (AFS). The AFS consolidated all BLM operations and centralized the statewide operations into the Fairbanks office. The move authorized the Alaska Fire Service to perform any suppression jobs for all the Interior agencies; permitted the partition of Alaska north and south, BLM and Alaska Department of Forestry, to continue; and encouraged efficiencies during what would evolve into a long budget drought. Like a body in shock, the BLM was pooling its lifeblood into its central organs, away from its limbs. In a sense, it was reincarnating the Alaska Fire Control Service for a new era.

That resolved the issue of who would execute the new order. The policy track was equally innovative. The pivotal institution was the Alaska Lands Use Council created by ANILCA. The council provided the aegis for a Fire Management Project Group that brought together representatives of Doyon Ltd. (for the Alaska Federation of Natives), Alaska Department of Fish and Game, Alaska Department of Natural Resources, National Park Service, Fish and Wildlife Service, Bureau of Land Management, Bureau of Indian Affairs, and the U.S. Forest Service through its Region 10 and Institute for Northern Forestry. They elected to fashion a modern fire management plan, a template, using a 31-million-acre patch of central Alaska called the Tanana/Minchumina region. It included Denali National Park, Nowitna National Wildlife Refuge, Native corporation lands, Native allotments, and a scattering of settlements from Fairbanks to Tanana. A representative landscape for a representative plan.

How to consolidate planning across a region with so many landowners? The solution was to divide the area into four zones, each of which mandated a particular response. *Critical protection* indicated all-out suppression, with elements of urban fire services—a choice suitable for the outskirts of towns, for example. *Full protection* mandated traditional wildland suppression in areas that threatened vital resources. *Modified action* involved aggressive initial attack "unless otherwise directed by the land manager/owner upon completion of a modified initial attack analysis." And *Limited action* required only that fires be contained "to the extent required to prevent undesirable escape."[17]

These four options were drawn onto maps under the direction of the landowners, who decided which policy was best for each site. The "critical" zone was the smallest; "limited," by far the largest. In essence, fires near roads and villages would be fought aggressively; fires outside the road net, would be watched and perhaps loose-herded or contained by burning out from a reasonably secure line like a river. There was no discussion of prescribed fire, the coming tool of choice in the Lower 48. Instead, fires free-burning in the limited zone would do that ecological work.

The Alaska Fire Service would execute whatever action the landowner determined was appropriate.

But parcels among landowners were scattered and remote. For each agency to meet its fire management goals separately would mean hopeless overlays of duplication, which meant alternatively that they would have to consolidate operations without surrendering control over policy. In 1979 the BLM and ADOF agreed, in the name of efficiency, to partition fire protection responsibilities between them along broad geographic zones, roughly midway east and west through the state. The state would handle all fires, regardless of ownership, in the south; the BLM, in the north. The consolidation would take place over several years.

By 1982 all the federal agencies had received legal title to their lands under ANILCA, and the Native corporations and the state had selected most of their lands. The *Alaska Interagency Fire Management Plan: Tanana/ Minchumina Planning Area* was ready for signatures in March 1982. On March 17, 1982, Secretarial Order No. 3077 (Wildfire Suppression and Fire Management–Alaska) granted the BLM authority to function as a fire-service organization for the Department of the Interior and Native corporative lands (the Native lands were to be considered "on an equal basis" with federal lands). The Alaska Fire Service remained under the BLM Alaska. By December 1, 1982, the federal agencies were to encode the new order into an updated departmental manual. Each agency would contribute "pre-suppression and support monies commensurate with fire problems on their lands." The cost sharing did not extend to emergency suppression costs. On May 28, the BLM issued an instruction memorandum (82–226) that confirmed an Alaska Interagency Coordination Group as a mechanism to ensure cooperation among the federal agencies and Native corporations.[18]

It all seems the only strategy that might have succeeded, but of course there were many options that could have failed, and it was not foreordained that the Alaska model as we know it would exist. It probably helped that the U.S. Forest Service was a minor actor, left to its fire-intolerant coastal forests and a small research presence through the Institute for Northern Forestry; that the major players at the table were agencies under a common department (Interior) and the State of Alaska. It probably helped that 1981 and 1982 were light-load fire years nationally, which granted the agencies some bureaucratic breathing room. It helped that the Alaska

Fire Service had a preexistence outside the BLM, which made it easier to accept an arrangement sideways to BLM bureaucracy. It helped that a budget crisis imposed by the Reagan administration stripped away any option for separate agency fire programs. It helped that Alaska was not deeply burdened with the kind of fire-suppression pressures and history that made it difficult in the Lower 48 to move reform from position papers and memos into field operations. The political planets that had to align did. But then there were plenty of potholes and trolls along the path that could have stopped the project.

In retrospect, it seems that what evolved was the only thing that could have evolved. It wasn't. Any number of events or personalities or grievances would have derailed the project. But they didn't, and Alaska became among the luminous examples of progressive fire management in the country, one of the few unadulterated successes of the fire revolution.

What was consolidated in 1982 became the template for another dozen similar plans across the state. In 1998 all of these plans were merged into a common Alaska Interagency Wildland Fire Management Plan, formally incorporating the 1995 common Federal Wildland Fire Policy guidelines. (That year the Carla Lake fire near Delta Junction burned 54,000 acres, mustering 52 crews, 32 engines, 622 overhead, and a $10 million suppression bill.) The plan was again updated in 2010, after the horrendous 2003 and 2004 fire seasons, and once more in 2015, following another near-record outbreak. Despite tweaks, upgrades, bureaucratic hacks, and the chronic instability of government at both the state and federal level, the Alaska model has thrived for 35 years. With further amendments it might survive another 35.

The issues it confronts are legion, as abundant as in the Lower 48, but with an Alaskan accent. The workforce, particularly Native crews, is declining due to drug and fitness testing, alternative sources of income, and the unreliable demands of each season. Alaska used to assemble hundreds of emergency fire fighters, not only for internal use, but as a national resource available for export. Now numbers are a fourth of what they were when the Alaskan model jelled. The wildland-urban interface is growing. Mostly this is an Alaska Department of Forestry responsibility, but it affects how the duopoly that manages wildland fire in Alaska operates. The Alaska Fire Service has a quirky equivalent in the thousands of Native allotments (and hundreds of villages) it must protect. In Alaska, as elsewhere, attention paid to the wildland-urban interface is attention and monies not devoted to other tasks. Budgets are not only shriveling but are increasingly unpredictable. Federal paralysis has compelled the Alaska Fire Service to rethink how much to centralize in the name of cost savings and how much it might reopen outlying bases in the name of policy effectiveness. The oil glut created by fracking threatens the State of Alaska with a fiscal crisis that has rippled through its capacity to hold the workforce it needs and left it to mutter about renegotiating the grand partition of the state between it and the Alaska Fire Service (southwest Alaska is an obvious point of contention). The Native corporations are contracting for carbon credits, which will bring in revenue but leaves the Alaska Fire Service wondering at what point the corporations have acquired the "substantial revenue" that will allow them to take over fire protection chores. Even among federal cooperators there are concerns over who is doing what; the 2000 National Fire Plan boosted the fire staffs of the National Park Service and the Fish and Wildlife Service, which left the Alaska Fire Service worried about overlapping duties. And there is the existential challenge, the great unknowable, posed by climate change and associated ecological upheavals. Demands are growing; support is declining.

Yet such concerns are grit in the gears that regular maintenance can clean away. They are manageable so long as the basic apparatus endures and the partners believe that the present modus operandi is the only meaningful modus available. All parties appreciate that, if dissolved, the arrangements are not likely to be reconstituted. The Alaska model was the outcome of a peculiar pyropolitical moment. No one in 1967 would have predicted something like an Alaskan model 15 years later. No one today can declare that a comparable surprise might not happen by 2032.

What does remain, though, is speculation about how the Alaskan model might be extrapolated outside the Last Frontier. After all, interagency plans and operations have become the norm; restoring fire on a landscape scale is a national goal. Plans are floated almost yearly to transfer at least large-scale fire suppression to FEMA. The Alaskan model has achieved what no other region has. Why shouldn't it head south, like sourdoughs with their pokes filled with gold?

It's a good question, certainly one that Alaskan fire officers are not afraid to voice. The simplest response is that the model is unlikely to work because it was the improbable outcome of special circumstances—a historical accident, in the truest sense. It was not designed from MBA-taught first principles to meet management needs. It evolved in quirky ways in response to a cluster of pushes and pulls unique to Alaska around 1980. The Florida model has not translated well outside the Southeast despite what would seem to be the transcendent logic of prescribed burning. The Southern California model has not exported much beyond California. The United States has many models for fire management based not just on fire's ecology and behavior but on the oddities of landscape, fire culture, historical opportunity, and politics. There seems no logical reason why what works in one place shouldn't work in another. They just don't, or don't unless they come with major adaptations.

During the Alaska wilderness debate, an oft-cited factoid was the people farthest from wilderness valued it most. So, too, the Alaska Fire Service model can seem more attractive the further away the viewer is from the vexing hassle of managing the relationships that make it possible, from the relentless mosquito-and-black-fly issues that plague daily operations, that see a universal, first-principle model, not something constructed from historical circumstances. The Alaska model is not a prepackaged kit on a pallet that can be paracargoed and dropped elsewhere like a radio system. It more resembles a village. There are subregions where something like the Alaska model is possible, and may happen, in the same way that there are patches of the country outside the Southeast that have succeeded with prescribed fire as a foundational practice and places where something like the urban-fire-service-in-the-woods model of California makes sense. But it is unlikely to serve as a national exemplar.

Of all the historical accidents that made the system work perhaps the core was the creation of the Alaska Fire Control Service. It was always an improbable invention, with a fireguard or two riding speeders along the Alaska Railroad between Seward and Fairbanks. It's hard to imagine a federal government other than one like the New Deal conceiving it. But it survived after a fashion because of the imminent war. Although the BLM absorbed it as it did the General Land Office and Grazing Service, it was never fully dissolved and reconstituted. It resembled the mitochondria in a cell, one organism absorbed by another, separate but symbiotic. It could keep its culture, the traditions that gave it an identity. It could later reincarnate itself as the Alaska Fire Service.

Because the Alaska Interagency Fire Management Plan left policy and management goals with the landowners, the Alaska Fire Service could do what it had mostly done over 40 years of history. It could suppress fire, or where agencies wanted more fire on the land, it could direct big burns away from critical sites and let them ramble through the bush. Natural fires could do what elsewhere required prescribed fires. Unlike the National Park Service in the Lower 48, the agency did not have to live with two internal fire organizations, one dedicated to suppression and one to prescribed fire, like Roman tribunes governing an army on alternate days. It could act with consistent force when called upon.

The freak of history that made the Alaska model possible is probably unreproducible at scale. But then the same could be said of Alaska.

THE ALASKA FIRE SERVICE

THE ALASKA FIRE SERVICE has its headquarters off the runway at Fort Wainwright, outside Fairbanks. The original structure was erected during World War II to support the Lend Lease program that sent American airplanes to the Soviet Union; Fort Wainwright was where American pilots turned over the keys to their Soviet counterparts. The two historic western outbases of the Alaska Fire Service, Galena and McGrath, were alternate fueling sites for the flight to Nome and then across the Bering Strait. Like much of fire management in Alaska, an infrastructure created for one purpose has evolved to support newer ends. What endures is aircraft, and an origin as a logistical hub.

There is not much option. Most of the larger towns and villages reside along rivers (salmon is a vital food source), but barges are slow. The real docking point with the outside world is the airstrip. The immensity of the lands under protection explains why. The Upper Yukon zone spans 50 million acres, roughly the size of Louisiana, and has a collective population of 1,800, approximately a tenth the size of Natchitoches City. The Tanana zone embraces 44 million acres (think Missouri); its entrepôt, Tanana, has a population of 245. The Galena zone sprawls across 93 million acres, roughly the size of Montana, and includes such metropoli as Kotzebue (3,284) and Nome (3,788).

For a mission that involves wildland fire across such expanses the only hope to respond in a meaningful time is aircraft. For aircraft to work

requires an elaborate complex of supporting infrastructure and institutions. During the great build out of the 1950s and 1960s, when the BLM
had responsibility across the state, outstations were established to improve
initial attack, and those outposts had their own outstations. Since the succession of partitions, however, the BLM has consolidated its operations
into the Alaska Fire Service, and it began a great build-down, a slow
implosion of infrastructure that has concentrated more and more functions in Fairbanks. Now an awkward balance exists between the efficiency
gained by centralizing operations and the effectiveness required to protect
mines, remote fishing camps, hunting cabins, and village allotments.

Both demands translate into logistics. The Alaska Fire Service operates the largest aerial fire program in the United States. It runs the largest
civilian paracargo operation in the world.

———————————

Mike Roos, assistant fire management officer for the Galena zone, likes
to make things happen. A sense of adventure took him to Alaska, a fasci-
nation with aviation led him to the Alaska Fire Service, a taste for action
brought him to logistics. As the adage goes, young generals talk strategy,
old generals talk logistics. Mike Roos is an old general in Alaskan fire.

He was born in Chicago in 1958. In 1976, wanting to go West, he
enrolled at the University of Alaska-Fairbanks. By 1979 he was working
aboard a crab processor in the Aleutians. That year he applied to the
Alaska Department of Forestry, then flush with oil money and anxious
to reclaim fire protection from the Alaska Fire Service. The state was
developing agricultural lands around Delta Junction, and ADOF had to
deal with immense slash piles and escapes from the resulting burning. In
1980, just as Alaskan fire was rebooting, Mike was hired. His entree into
fire whetted his enthusiasm for matters aerial. The Alaska Department of
Forestry had a helicopter crew. The Alaska Fire Service had smokejumpers, paracargo, and aerial attack.

Over his career he moved between those two institutional polarities
of Alaskan fire. In 1986 he transferred into the Alaska Fire Service as a
smokejumper. He returned to ADOF in 1990 and worked in a variety of
aviation and logistics roles, ultimately becoming fire management officer

for southwest Alaska, which was then managed through an outbase at McGrath. In 2006 he recycled back to the Alaska Fire Service in operations. In 2013 he swung back to ADOF as fire officer for McGrath to complete the years he needed for formal retirement in Alaska public service. In 2016, retired from ADOF, he returned to the Alaska Fire Service as assistant fire management officer for Galena. Along the way he joined the Marine Corps reserve, mostly attracted to reconnaissance ops.

During his 35-year career he had done nearly every fire job available. He saw the fire scene from both perspectives, state and federal (he decided that most of the rivalry originated with the state, not the feds.) Mostly, though, he was drawn to the aerial operations. He decided he liked the remote fires, not the roaded ones. He liked helitack, smokejumping, para-cargo, air attack supervision. He liked the western landscapes of Alaska. He was eager to reposition himself in Galena.

⸻

I asked him for an example of logistics at work. He described a hypothetical fire (not so different from a real one in 2016) that burned near Dahl Creek in the Galena Zone, located near Kobuk, equally distant east from Kotzebue and north from Galena. Dahl Creek was once the site of the northernmost fireguard station in the United States. It is now a wooden sign hanging in the warehouse at Galena.

A smoke was reported by an aerial observer, flying reconnaissance after a lightning storm passed through the area. The lightning-detection network had recorded many ground strikes. With better resolution, remote-sensing satellites like MODIS might have picked it up.

Because it was near Kobuk, the fire lay in a full protection zone. The village and Native allotments needed shielding. Eight smokejumpers flew out of Galena in a Casa 212 jumpship. Equipment, radios, food and water for three days, "wrapped and strapped" with a cargo chute attached, were onboard the Casa and parachuted in with the jumpers—enough for three days.

The fire burned in black spruce, and without any humidity recovery over the evening—there was no evening, just a full-day burning period for a "day" that could persist for weeks—the flames didn't pause but fed back into their own momentum. The fire bolted beyond control. A 4-acre initial-attack fire morphed into a 100-acre extended attack, then to 500 acres. A fire behavior forecast predicted more of the same.

This required additional jumpers, and the Alaska Fire Service dispatched two more jumpships. Air tankers of several types responded along with aerial supervision from Fairbanks. CL-415 water scoopers, small Fire Boss scoopers, along with Convair 580 retardant tankers soon operated overhead. Two emergency firefighter crews were called up, along with the Midnight Sun Hotshots. They flew directly into the Dahl Creek airstrip in Beech 1900s and Cessna Caravans. A Bell 212 and a Bell 407 helicopter responded out of Galena; fire specialists set up a helibase and staging area. AFS aviation provided fuelers and flew in their equipment. Even a historic workhorse, a Curtiss C-46 that had flown over the "Hump" in World War II, arrived with a load of jet fuel to sustain local air operations.

What had been a sleepy airstrip above the Arctic Circle now buzzed with activity, and that activity had to be managed. An air tactical group supervisor oversaw the air space. A radio net was ordered, flown in, and set up. A field camp sprang up suitable for 80 firefighters, camp workers, and overhead.

Crews labored to create fuelbreaks and defensible space around Kobuk and nearby allotments, then burned out into the bush, running pumps and hoses to hold the line. The Bell 407 widened the burnout to the Kobuk River by depositing incendiaries from a Premo PSD (plastic sphere dispenser) that injected ping-pong balls with potassium permanganate and ethylene glycol, each ball erupting into flame after being dropped.

The scene began to resemble a project fire, maybe a campaign fire. More supplies had to be flown in. The helos burned through their fuel. Sometimes commercial airstrips can serve as depots, particularly for aviation gas, but Dahl Creek was as close to Galena as to Kotzebue, so Galena remained the primary staging area. The Dahl Creek fire was burning through some serious woods. The Alaska Fire Service was burning through serious cash.

On it goes. The fire would rage through as much black spruce as it could reach by surface flame or firebrand. With the critical sites shielded by breaks and blacklines, the Type III incident management team worked to secure the near-fire perimeter while allowing far-spreading front to move on, now splashing over 40,000 acres. The flames were left to probe, push, pull back, pulse, and expire as they would. Aerial observers watched and monitored.

The main fire organization now went into reverse. Demobe was no less complicated than buildup, just less frenzied—or only slightly so since materiel and crews no longer needed at Dahl Creek would likely be needed somewhere, and would be redirected through Galena. Everything brought in had to be brought out. All the crews. All the pulaskis, pumps, and swatters. All the food and water. All the camp equipment. All the garbage. All the fuel. And all of it had to be removed by air. The supply chain, like a mechanical caterpillar, bunched and stretched itself back to Galena, and then to Fairbanks.

In the end, the Dahl Creek fire was one of 56 reported. The Alaska Fire Service had taken similar action on 19 of them. One exceeded 50,000 acres. One was 128,000 acres and still creeping and sweeping across the Alaskan bush. It would burn until the snows snuffed it out.

Logistics is everything.

In the early days, the AFCS would have taken action on every fire it could reach. Those that escaped and got big—the genies that slipped out of their bottles—it would have to leave; but policy and temperament dictated an attack on every fire. It was a kind of sublime madness. With statehood and ANCSA the BLM began yielding protection responsibility to a fledgling Department of Forestry. Following ANILCA, policy changed because the bulk of lands, and those deemed most valuable, transferred to the Fish and Wildlife Service and the National Park Service, both of whom wanted more free-burning fire on the land. Mass suppression shrank to the protection of points, most of which fronted rivers and coasts. Even as the technology for fire management expanded exponentially, the arena to use it contracted.

For a while, logistics was limited only by the willingness of fire officers to call on it. Now it's limited by the willed values of landowners. In the past the limits of what you could do with fire in Alaska depended on the capacity of your logistical operations. Now logistics is constrained by policy. We have the capacity to do more. We just choose to do less.

Even so, the demands for protection remain, and because the fire scene is so diffuse, it will remain committed to aerial response. The firescape is changing. Policy, or its interpretation, will surely change also. But whatever is done under whatever guidelines, it will be done by air. And it is likely the Alaska Fire Service will do it.

LAST FRONTIER OF THE U.S. FOREST SERVICE

WHAT HAPPENED TO the Forest Service?

One of the political paradoxes of Alaskan fire history is that the federal agency that first addressed the fire issue a century ago has today the least responsibility for managing it. It was the U.S. Forest Service that first brought a vestige of formal fire control to the state, that campaigned for fire protection in the interior, that transferred personnel to staff the Alaska Fire Control Service, that conducted the important research in fire science, and that now has a negligible presence on the Alaskan fire scene.

The Forest Service lost its fight with Ballinger, then lost even bigger with ANILCA. All its historic rivals got land, power, and visibility. The Forest Service was left with costly legacy entities like the Institute for Northern Forestry that it could not afford and with festering issues like logging in the Tongass that it could not resolve. It has the least pyrophytic landscapes in Alaska. It has the least say in how fire management evolves. It's simple, though inaccurate, to project the present back into the past because the projection has to pass through a camera obscura that inverts the narrative.

The early forest reserves, imagined and inspected while they were still under the General Land Office, were relatively incombustible. They were extensions of the Pacific Northwest coast, and concerns were over logging rather than burning. The most vulnerable was the Chugach, which extended into the Kenai Peninsula. Here, local slashing and mining created pockets susceptible to fire when the north winds, a local foehn, blew. W. A. Langille observed that there was a "period in the early summer of each year when the prevailing north winds dry the surface of the tundra and forest mosses to such an extent that they readily ignite, and once caught, fire spreads rapidly. . . . Where burned, every living thing, even to the heavy sphagnum mosses was killed . . . [and] *not a single spruce seedling was seen.*" That prospect remained more fear than practical threat.[1]

The real concern was the interior. Investigating Alaskan forests in 1909 assistant forester Royal S. Kellogg reported what every other progressive observer did, that while cutting remained local, fires were widespread. "It probably would not be far from the truth to say that in the Fairbanks district ten times as much timber has been killed by fire as has been cut for either fuel or lumber." Fire followed the white newcomers, for all the usual reasons. "During the entire trip of 460 miles down the river from Whitehorse to Dawson," Kellogg thundered, "one is almost constantly in sight of fire-killed forests." The locals considered that mosquitoes were the greatest cause of wildfires since smudge fires (many of which escaped) were everywhere. The fire-immune coastal forests were reserved before they were greatly "impaired." The interior forests were subject to unregulated cutting and unremitting burning. "Their protection can not begin too soon."[2]

Others followed—James B. Adams and Earle Clapp in 1913, Henry Graves and E. A. Sherman in 1915, Arthur Ringland to Kenai in 1916. Chief Forester Graves had a keen eye for fires as well as timber. Fires around the Kenai and an embryonic Anchorage followed the Alaska Road Commission and the new railroads; the fire scene in the interior was unspeakable. "The interior forests of Alaska are being destroyed at an appalling rate by forest fires. Conditions existing in the western United States 25 years ago are repeating themselves in Alaska. The entrance of the white man brought the forest fire, and he has succeeded in a short period of less than 20 years in destroying the forests to an average extent of fully a million acres a year." The General Land Office did nothing to stop those fires. Graves thought a fire-protection system like California's might be appropriate. In 1922 John Guthrie repeated Graves's observations and furthered his argument. Some 25 million acres had burned since the white man arrived; many millions have burned "over two or three times leaving utter waste." No agency, "governmental, territorial or private," dealt with fire. Yet, "as a national duty, it is imperative" that the federal government, the primary landowner, "protect Alaska's forests against fire," and as the agency best equipped by law and experience to handle fire, the Forest Service would seem the "logical" choice. Guthrie did not argue for putting the interior forests into national forests, only that "protection from fire should be delegated to the Government service whose special function is the protection and administration of Federal forest lands." What the Forest Service could not (or would not) do on its own, it would assist others to do. The absence of organized fire protection was an institutional vacuum it abhorred.[3]

It took a visit from Secretary of the Interior Harold Ickes in 1938 to begin a formal response. The vast burning shocked Ickes, who consulted with Ernest Gruening and Frank Heintzleman (supervisor of USFS Alaskan forests). The upshot was legislation enacted by Congress to establish the Alaska Fire Control Service. The Forest Service not only extended further advice to the embryonic agency, it allowed W. J. McDonald, supervisor of the Chugach National Forest, to transfer to lead it, along with others trained in Montana. McDonald brought in Harold Lutz, a young graduate of Yale forestry, later the author of two formative studies published in the 1950s that deeply influenced thinking about the nature of Alaskan fire. When the AFCS was stirred into the bureaucratic stew that became the BLM, another Forest Service fire officer, Roger Robinson, accepted appointment as its director. When big fires blew over the interior in 1957, the Forest Service dispatched Charles Hardy from the Missoula Lab to report.[4]

That so many Forest Service personnel transferred to the BLM helps explain why it emulated Forest Service ambitions so thoroughly. Until the agency could build up its capabilities, the Forest Service filled in, a liaison strengthened with statehood. The McSweeney-McNary Act was extended to Alaska while it was still a territory, allowing for Forest Service research to establish itself in time to create Alaska's first fire-danger rating system. In 1960 Robinson explained in the *Journal of Forestry* how modern—that is, aerial—fire control had come to Alaska. The Clarke-McNary program extended federal cooperation to the state forester, along with assistance from the Forest Products Laboratory to the mill in Wasilla. The agency then set up a Forestry Sciences Laboratory at the University of Alaska. In 1963 it established the Bonanza Creek Experimental Forest near Fairbanks and published a survey of the Alaskan fire scene. In brief, where gaps remained in a comprehensive fire program, the Forest Service plugged them.[5]

After the 1969 Swanson River fire, the BLM assumed more of the core tasks, including some exploratory science and hosting symposia. But Alaska also felt the revolutionary rumblings from the Lower 48. As a bureau, the BLM was ascending, and the Forest Service entering a painful decline. Outside its own lands the USFS would find it difficult to invest much money and energy; the Institute for Northern Forestry looked like the Institute of Tropical Forestry it maintained in Puerto Rico. By the time it completed its internal fire reforms in 1978, it was poised for a withdrawal. Among federal land agencies, it was the big loser in ANILCA. The national parks doubled in size; so did the national wildlife refuge system; the National Wilderness Preservation System tripled. The national forests got nothing. By 1997 the Institute for Northern Forestry had so shrunk that it was simply shuttered. The Bonanza Creek Experimental Forest segued into a long-term ecological research site. The Forest Service remained on statewide fire planning teams, but without many fires, and with a robust institutional matrix (even a quiet rivalry between the Alaska Fire Service and the Alaska Department of Forestry), its presence was more ex officio than exemplary.

There is an argument, too, that having the Forest Service at arm's length meant that Alaska did not suffer through the oft-agonizing crises that accompanied the Forest Service in the Lower 48. The 1980 reboot quickly pushed the Alaskan fire community into modern fire management. The 1980 elections set the Forest Service on the path to paralysis. What happened to the Forest Service would stay with the Forest Service, and out of Alaska.

A wider vision might note that America is good at startups (and gold rushes), poor at maintenance (and picking up the messes it makes). It's good with young adulthood full of passion and promise, poor at middle-aging with its complications and legacies. The 1980 reforms in Alaska made a new startup called the Alaska Fire Service. The 1978–80 reforms in the Lower 48 left the Forest Service to grapple with a kind of midlife crisis, with the many responsibilities of a mature agency that could no longer move as nimbly as in its past and was shackled by controversies it had no power to resolve.

The Forest Service had helped birth fire protection in Alaska and sustained it through its adolescence. Then, like many a parent, it passed to the sidelines.

LIVE-FIRE ZONE

THE ALASKA FIRE SERVICE fights fire. That's what it has always done, and so long as it survives, what it is likely destined to do. Where fire needs to be restored, AFS and the land agencies defer that task to nature by leaving lightning ignitions room to roam. Prescribed fire is not a goal, nor prescribed crown fires, an option, despite a handful of experiments for wildlife habitat. The primary firescape, black spruce, doesn't underburn: it either simmers or boils over. Hardwoods like alder, birch, and aspen don't burn briskly in most circumstances, and serve as de facto fuelbreaks. The uncertainties over burning tundra are too great to warrant introducing anthropogenic fire at any scale. Yet in one anomalous arena prescribed fire is the treatment of choice.[1]

This is the 1.4-million-acre military zone, where the Alaska Fire Service manages fire on select Department of Defense bases in Alaska. Most facilities handle their fires internally. But there are training grounds, notably outside Fort Wainwright and on the Yukon Training Area, that are breeding grounds for fires started by summer exercises and live-fire munitions. Small arms, with tracer bullets. Howitzers. Striker vehicles. Hot casings ejected from Apache attack helicopters. Joint air and ground exercises with explosives. They can start lots of fires in places receptive to burning.

The land is co-managed by the BLM and DOD—this is public land used for military purposes under Public Law 106.65. It's premier training ground; few landscapes offer such bounteous openness; fewer come with equally abundant air space. Training costs per year average a billion dollars. The military needs to keep its fires (and any that start from other causes) from interfering with training and to keep wildfires from bolting out of the bases. Because of long, heavy use, much of the landscape is grassy. The obvious solution is prescribed spring burning.

The BLM and the army are the jurisdictional agencies. The Alaska Fire Service is the operational agency, charged with executing the fire job. On average it burns 65,000 acres a year with 100–150 fires. And it suppresses fires of any source that blast out of their containment lines or threaten to leave the bases, or fires that might burn into the bases from outside. As the complexity of the task has scaled up, so has AFS staffing. The military zone became one of four AFS management zones in the early 1990s. That brought a fire management officer, an assistant fire officer, a fuels specialist—by 2017, a staff of eight.

The program is evolving into a year-round operation. Spring burning occurs when the ground is frozen and the spruce still damp. Crews can use snowbanks as firelines. They blackline the target sites early, using drip torches and terratorches, then return later for heavy-acre production burns by aerial ignition. Summer and fall are times for wildfire—keeping fire out of training sites, keeping outside fires out, and inside fires in. The late fall and winter offer occasions for fuel treatments, rolling back black spruce where training demands more ground, piling and burning slash.

For decades a simple barter system paid for the arrangement. The AFS kept its facilities complex at Fort Wainwright in return for its fire services. But in recent years, the economics of the relationship has become unbalanced. In 2014 a munitions fire, the Stuart Creek 2, overwintered, then blew up in the spring with a $25 million suppression bill. In May 2014 the 100 Mile Creek prescribed fire in the Oklahoma impact area escaped, spotted into an unexploded ordnance area, which prevented suppression, and went wild over 23,270 acres at a cost of $20 million. The Alaska Fire Service calculates its rent at approximately $2 million a year. In the past two years suppression expenses, absorbed by the AFS, have run to $40 million. A new agreement has each side billing the other for costs.

——————

Throughout its lands the Department of Defense starts lots of wild fires, arranges for lots of prescribed fires, and generally takes fire seriously. Eglin

Air Force Base in Florida prescribe-burns 100,000 acres a year. The artillery range at Fort Sill starts fires year-round that sustain some of the finest prairie in the Great Plains. Camp Pendleton is so laced with fuelbreaks that an aerial view resembles the furrowed exterior of a grenade. Vandenberg Air Force Base had, for 30 years, its own hotshot crew. But Alaska of course is different.

There is something agreeably Alaskan about the migration of big game that pass through the postburn sites. Moose drift across wetlands. A 350-strong herd of bison managed by the Alaskan Department of Fish and Game wander between state and DOD lands, enjoying the fresh greenery that follows spring burns. The origin of the fire may matter less than the habitat it leaves behind. The ownership of the land may matter less than the fact that it burns.

SPARKS OF IMAGINATION

EVERY REGION HAS its repertoire of stories that illustrate how little the outside world knows about its fire scene. In Alaska that historical howler revolves around the ignorance of early whites regarding lightning as an ignition source. They thought most fires were started by people. They even published peer-reviewed monographs that elaborated on the subject. They didn't know.

Telling the story is a badge of membership in the Alaskan fire fraternity. But such rituals always come with a policy subtext: they help explain what we should do (or not do). If people start fires, you organize your suppression organization around one set of practices. If lightning, however, kindles the problem fires, then you organize around another set. The rituals, too, also come with a moral subtext. In this case, the emphasis on human ignition says that Alaska's problem fires were the product of bad behavior. They were of a piece with overhunting moose, overfishing salmon, hydraulic mining in the gold districts, logging in ways that (in Teddy Roosevelt's phrasing) "scalped" the land. What humans did humans could undo. What nature does, however, is not easily or wisely undone.

———

There is no reason to think that the early observers reported inaccurately. They were credentialed scientists, often in government service. Robert Bell, Alfred Brooks, and Harold Lutz were as credible as anyone in literate culture. They reported what they had personally seen, and where they scoured the literature for references, they documented their sources with full academic rigor. Of course they were products of their day and training, but if we dismiss them for those reasons we should dismiss the scientific community of today for similar cause.

They saw what they saw, but what they saw depended on when and where they went. They traveled along traditional routes, they had guides, they used railways and steamboats. They went where most people did, at a time when alternatives to open flame were few, and that meant they saw lots of fires started by people. Some starts were careless, some deliberate, some bolted beyond campfire or smudge. What Hudson Stuck wrote in 1917 could stand for dozens of commentaries: "Should the season be a dry one, the traveller is almost certain to encounter them [forest fires] somewhere along the course of the Yukon, and at times the journey down the river is made an almost continuous evidence of their activity, near and remote. Sometimes the whole river reeks with smoke from Whitehorse to Anvik."[1]

Such accounts of fires are nearly interchangeable with those from the frontier on the Lower 48—the Willamette and Sacramento Rivers and even the Great Lakes were frequently smoked in—and why shouldn't they be since the same kinds of peoples were doing the same kinds of things? The outcomes in Alaska differed because of the peculiarities of the boreal biota, but there is every reason to believe what the early observers wrote. It would be odd—would require explanation, in fact—if Alaskans had behaved differently.

The most copious survey is Harold Lutz's *Aboriginal Man and White Man as Historical Causes of Fires in the Boreal Forest, with Particular Reference to Alaska*, published by the Yale University School of Forestry in 1959. Lutz was then Jesup Professor of Silviculture at Yale, but his Alaskan expertise had begun earlier when, as a graduate student, he began research at the request of the Forest Service and AFCS. He spent many summers in the field; he spent long winter hours in the library panning through historical accounts.

It's worth reviewing his conclusions. First, on the indigenes, whose greatest source of ignition was campfires used for many purposes, and rarely extinguished. They were set for heat, light, cooking, warming the gum used to repair birch bark canoes, signaling. A few fires were set for hunting, and occasionally for warfare, but not routinely. Here and there fires were set to clear forests, fell trees, cut up logs, and kill woods to supply fuelwood. Ubiquitous, however, were fires whose smoke was intended to drive off mosquitoes and gnats. Where fuel, drought, and wind mingled, those fires would blast out into the woods. "It seems certain that even prior to contact with white man, aboriginal man was responsible for frequent and widespread fires in the boreal forest." Then came the whites, who were "without doubt" the cause of even more fires. They were careless and had the capacity to start fire easily. They abandoned campfires, set fires in live forests to create dead fuelwood, kindled fires to ward off insects, set signal fires, practiced some fire hunting, burned to promote grass for livestock, fired widely to expose the ground for mining, and "set the forest afire just to see it burn or 'for fun.'" Until well into the 20th century industrial combustion, or alternatives to open flame, were rare.[2]

Lutz acknowledged lightning as "certainly" one of the causes of fires, but his reckoning was that humans were a far "more prolific source." He was right. He remains right today. From 1956 to 2000 people set 62 percent of all recorded fires.[3]

Not until fire protection began to pursue fires into the backcountry did the magnitude of the lightning fire spectacle become obvious. From 1956 to 2000 lightning accounted for 90 percent of burned area.[4]

These were fires off routes of travel and away from patches routinely burned for traplines or berries or fodder. The human-caused fires could be reached by foot, canoe, steamboat, or rail. The lightning-caused fires could be addressed only by aircraft. Not surprisingly, the shift away from human fires occurred with the appearance of formal fire protection. The big burns that bothered the military by smothering the sky, that blasted across landscapes open to selection by the state and Native corporations, and that caused heart-stopping costs to suppress were in some respects an issue only when Alaska achieved statehood. By now industrial combustion in the form of internal combustion engines could counter open flame.

Within 15 years of statehood, however, another consideration shouldered to the fore. The fire revolution countered fire-as-bad propaganda by demonstrating that fire had been on Earth as long as terrestrial life. It was natural, and human use could mimic that naturalness through prescribed burning. More generally, the era was one that was eager to counter the plastic with the pristine, for which a wilderness preservation system was a supreme expression. A lightning-kindled fire bust in wilderness was the pyric equivalent of a wolf pack restored. Lightning fires were neither an anomaly nor a quirk, but the essence of fire as a natural process.

The first response to the discovery of numerous lightning fires by the BLM was to counter multiple starts with multiple initial attacks. This proved hopeless, so the emphasis shifted to a form of prevention. Smokey Bear couldn't help, but perhaps weather modification could. During the 1960s Alaska experimented with efforts to induce storms over drought regions and to suppress lightning in the clouds. (Much as the BLM established a smokejumper program by loans from Missoula, so lightning suppression built on Project Skyfire research out of the new Missoula fire lab.)[5]

The second response was to accept lightning as inevitable, and necessary. Lightning was a way to turn Alaska's size and remoteness to advantage in the national ambition to restore fire to wildlands. Such fires needed to be managed, not extinguished. The Alaskan Interagency Fire Management Plan allowed lightning to do what in the Lower 48 the driptorch did. By this time all those accounts about the prominence of human ignition seemed not only empirically misplaced but philosophically misguided. That old-timers did not recognize the prevalence of lightning became a badge of their ecological callowness. Everyone knew that the Alaska fire scene was about lightning fires.

Except it wasn't. Humans still set far more fires, and most of the fires that threaten the greatest social values; they just cluster where there are humans, and these days they can be extinguished relatively quickly and efficiently with the machinery of industrial combustion.

―――――――

Today, an ignition map of Alaska shows two stark patterns. One tracks, with astonishing fidelity, the road and rail network and the distribution of villages, exurban enclaves, and hunting and fishing camps. The other fills the intermountain region with a scatter that, over the decades, has saturated the interior with burns. There are clusters of lightning-fire busts by year, but the compounding effect, year after year, is to blanket the region with starts and burn scars. The two maps trace incommensurable, if not quite immiscible, realms of combustion.

And they define the two realms of interest by fire institutions. The state handles mostly human ignitions, which threaten life and property. The Alaska Fire Service handles mostly lightning ignitions. Because of the grand partition of Alaskan fire services north and south, each must cope with both kinds of fires, but the general division holds. The biggest exception is southwest Alaska, a state responsibility, though it is likely that the terms of partition will be renegotiated in the future.

Wildland and urban fire services are distinct cultures. They have fused, after a fashion, at great expense, in California, and Colorado is experimenting with a CalFire-lite version. But their goals, their character, their literature and art, their sense of themselves is distinct. In most of the United States they exist in segregated realms. That will likely be the future in Alaska as well unless settlement expands rashly into the bush, climate change blurs old borders, and fire ecology and fire economics, which presently support one another in Alaska, come into sharper conflict. If dispersed settlement worsens, it will complicate natural fire by demanding protection for tiny allotments, cabins, and lodges, and there will be complaints about smoke. What are points will morph into patches and patches into protectorates defined by how humans choose to live on the land.

Early interest focused on human fires; recent interest, on lightning fires. The arc may shift back, as it is doing in the Lower 48. Half of the 21st-century's notorious megafires have been started by people; human starts and invasive grasses are setting up an unholy alliance that threatens to unhinge whole ecosystems such as chaparral and sage steppes; the Anthropocene has concentrated attention on that spinning whetstone where people put an edge on nature. A renegotiation of circumstances and enthusiasms may cause future Alaskans to chortle at those benighted predecessors who thought that Alaska's fire problems were largely about managing lightning ignitions.[6]

IN THE BLACK

THE INTERIOR FOREST that mesmerized the early proponents of fire protection is far from singular, or continuous, or uniformly combustible. The conifers burn more readily than the oddly named hardwoods such as aspen, birch, and poplar, the south slopes more easily than north ones. When people speak of the savage burning of Alaskan forest that can overwhelm all efforts at containment, they are talking about one variant of this complex assemblage. They are speaking of black spruce.

═══════════

The interior has wetlands and uplands, valleys and mountains, north slopes and south; in floodplains, it fills with oxbows and lunate arcs of soils; and on the mountains, taiga can yield to tundra. The forest understory is rich with lichens, mosses, and shrubs, quickly saturated and just as rapidly leached of moisture. Fires compete with floods and frost as disturbances. There are edges everywhere; the scene goes beyond the status of a mosaic to that of an ecological kaleidoscope. All of the biota can burn under the right conditions—grasses above frozen soil, peat after drought, deciduous woods if laden with enough surface litter. But the routine burning, the burning that makes the Alaskan fire scene what it is, occurs in black spruce.

The black spruce is a pyrophyte—of a type with jack pine and lodge-pole pine, given to explosive, stand-replacing burns, regenerating in its own ash. It favors north slopes over south, wet soils over dry, perma-frost over unfrozen patches. It has shallow roots, easily destroyed by fire. It grows with stubby branches, like a fasces of oily twigs. It burns as a dependent crown fire, relit continuously from the lichens and mosses and shrubs that send flame upward. An atlas of burned lands is by and large a cartography of black spruce, either by itself or intercalating with tundra, hardwoods, and white spruce. Canadian researchers have never found a black spruce forest that did not arise from a burn. As a recent survey of *Alaska's Changing Boreal Forest* puts it: "Black spruce ecosystems are born to burn."[1]

Remove black spruce, and the pyrogeography of Alaska would display regimes more akin to a montane woodland than to the volatile firescapes it is normally compared to. One reason is that Alaska's boreal forest, unlike Canada's or Eurasia's, lacks a significant pine component. What mixtures of conifers do elsewhere, black spruce does here. On the grand landscape burns that are Alaska's glory the gerrymandering black marks the zone of black spruce. It occupies an estimated 44 percent of interior Alaska.

The future of fire in Alaska thus depends on the future of *Picea mariana*. But that future will evolve out of a long past.

A paradox of Alaskan climate history is that the interior, as a part of Beringia, did not glaciate. It remained a grassy steppe, inviting access by fauna and humans. When the interior seriously warmed, around 13,000 years ago, it became suitable for trees. Over the next 3,000 years the region morphed from steppe into woodland, dominated by *Populus*, mostly aspen. Then, beginning 10,000 years ago, white spruce defined the biota for 5,000 years. After that era, over the past 5,000 years, black spruce has ruled.[2]

The prevailing fire regimes changed accordingly. In gross terms char-coal notched up during each phase change. The deciduous woodland phase burned, but fire had to compete with a full complement of megafauna browsers and grazers, and the prospects for large fires were scant. Without crowns there was no crown fire. The white spruce era had a steady background count of fires, but few conflagrations. With the transition to black spruce the overall charcoal load increased dramatically and spikes appear, indicating an alternating current of vigorous and languid fire years. With the final colonization by black spruce, the modern era commences. Counterintuitively, it may have been a shift to wetter conditions that allowed white spruce to invade steppe woodlands, and a further wetting that encouraged black spruce, which can thrive in wet soils, to replace white spruce, which requires a drier site. Warmer doesn't simply mean drier; warm air can hold more moisture than dry.

In the Alaskan boreal, moisture regimes can be further complicated by permafrost. Much of the requisite soil moisture is locked up in a peaty frozen ground, which releases water depending on the thinning and thickening of its surface insulation. What more than anything else controls that biotic insulation is fire. Severe burns—which means fires that linger into late summer and fall and tend to burn more deeply—can create a pyrogenic karst, or a fire equivalent to the prairie potholes left by receding ice sheets. That fire and water form a tangled skein is not news. What makes the circumpolar boreal, and especially Alaska, special is how frost acts as buffer between fire and water.

Still, the records—too sparse to be conclusive—suggest that fire return intervals have shrunk over the past 200 years. For upland forests an interval of 200 or more years seems plausible; today, it may be a third that. What is obvious is that we don't have data sufficient to reconstruct that past with any detail; that, for extended periods, regional differences can override general trends; that the upheavals set into motion by European contact, especially the gold rush, have erased an effective ecological baseline; and that, beyond the truism that every time is a time of change, the past may not be helpful in predicting the future.

In 1899 the Harriman Expedition—a sporting excursion financed by E. H. Harriman, director of the Union Pacific and Southern Pacific Railroads, that also carried scientists—visited Alaska's southeastern and southern coast (and Kodiak Island, where Harriman shot his trophy bear). The expedition's geomorphologist, G. K. Gilbert, commented that a change in climate, such as wetting and warming, could have opposite effects on the glaciers of the region so that most would recede but a few expand. A similar observation can apply to fire. Not every forecast has a trajectory that leads to more, or more feral, fire.

For most researchers fire's Alaskan future seems to be at a tipping point, teetering in the winds of climate change like a black spruce with its roots burned.

A warming climate is visibly melting the arctic ice pack. It seems to be accelerating burning in the interior and perhaps on tundra; the area burned between 2000 and 2009 is twice as large as any decade since 1940, when records begin. It also contains two of the three largest fire seasons on record, 2004 (6.6 million acres) and 2005 (4.6 million). The known fire return interval overall has shrunk from 196 years to 144. In 2007 the Anaktuvuk River fire more than doubled the known burned area of North Slope tundra. The fire season is lengthening; fires are shifting somewhat to late-season burning. Unlike the Lower 48, Alaska offers a clearer signal of climate change, less cluttered by legacy landscapes from a century of suppression and invasive species, less noisy with policy changes and fire officer choices. Still, scientific observations only date from the latter 19th century. From the Kennicott Expedition into the interior in 1865 to 2017 is 152 years, or 44 years less than the estimated fire return rate.[3]

The absence of a sure baseline compounds the uncertainties. It's not just European contact, or the pandemonium unleashed by the cascade of gold rushes in the late 19th century, it's also the long rhythms of the Milankovitch cycles, such midrhythms as the Medieval Warm Period and the Little Ice Age, and shorter cadences as the Pacific Decadal Oscillation and El Niño-Southern Oscillation. Until the past century Alaska's taiga and tundra had a relatively stable climate for 5,000 years, a box within which pieces might rattle and recombine but that held within certain boundaries. Over the past century those borders seem to be moving, with humanity as mover and shaker.

Wholesale changes in the biota are likely, and with them major restructurings of fire regimes. Those changes will be felt differently in tundra and taiga, wetland and upland, north of the Brooks Range and south of the Alaska Range, and throughout the kaleidoscopic intermountain boreal forest. Those biotic changes will rewrite fire regimes; in fact, fire may act as an ecological catalyst. The changes in fire will demand reforms in policy and institutions. Throughout there will be feedbacks; a slow burn of the tundra may be worse for carbon sequestration than flash burns across the boreal forest. But whether the shifts, or outright overturning, will lead to conflagrations will likely depend on the status of black spruce.

The fire history of interior Alaska is not really that old. When humans crossed Beringia there were still mastodons and wooly mammoths. Fire crews occasionally find tusks along their firelines (one hangs in the Fairbanks smokejumper loft). Alaskans have seen the landscape shape-shift from grassy steppe to woodland to white spruce to black spruce. The trend, allowing for bumps and blips, has been one of gradual warming and wetting. Future Alaskans will likely see more of the same.

The fire province that the Alaska Fire Control Service and its descendants knew may appear to future Alaskans as a fleeting one, whose spikes and troughs—so alarming at the time—appear trivial, almost stable. The future promises fire, perhaps announced through a slow wave of burning that remakes the scene. What expressions the subsequent, residual burning might assume is unclear. The Pyrocene may bring a menagerie of flames perhaps as strange compared to today's fires as Pleistocene mammoths and short-faced bears are to caribou and moose.

KENAI

THE KENAI PENINSULA IS a promontory, a beacon, and a portal. It's a geographic portal to the interior, a historical portal to Alaska's past and perhaps its future, an ecological portal to Alaskan fire. That's another way to say it's a transition point. Its eastern half is mountainous, its western a lowland plain. Its forest is a hybrid, both maritime and boreal. It's half wild and half semideveloped. It's as good a port of entry as any for a fire survey of Alaska.[1]

What may most characterize the Kenai, however, is its absence of a reliable baseline. There is no fixed point by which to measure the changes that are recorded. The climate has mutated continually, and recently seems to be passing through a major inflection point. The habitats have changed, and may be in the process of conversion. The species have changed; even the moose that were the reason for creating the refuge may be relative newcomers. The human history has changed, and is currently passing through a phase change not merely in demography but in the mode of settlement. The fires have changed; there are even reports of lightning fires, which were unknown previously.

All in all, the Kenai is famously active, but it's unclear whether that's because it has more happening or whether, being close to Anchorage and historic points of colonial contact, it has more records. What is clear, however, is that it has few obvious reference points. Apart from the grossest, geological markers there is no restore point for management, no anchor point for narrative. The Kenai offers a landscape expression of mindfulness. It's a continual now.

The Kenai compresses Alaskan pyrogeography and pyrohistory into a peninsula. Compared with most of Alaska, the Kenai fire story is known with some thoroughness. It has coastal forests, mixed interior forests, tundra, muskeg, and Alaska's three spruce, Sitka, white, and black, and in keeping with its transitional character, a hybrid, the Lutz. Sitka spruce burns rarely. White and black spruce burn as their understories and ambient conditions permit. What it hasn't had historically is the pyrotechnic busts that lightning kindles in the interior.

Charcoal cores extend back 13,000 years. They record three long waves of biomes and associated fire regimes. An early tundra era had the longest return interval; a woodland era shortened it; a black spruce era, surprisingly, lengthened it again. Over the past 300 years, which is the age of the oldest spruce, research suggests an average fire return interval of 89 years, plus or minus 43 years. For black spruce the interval is 79 years, for white spruce 200-plus years, and for mixed spruce forest 170 years. Very few sites in Alaska can boast of such detail. The record spans the entire contact era, predating even Bering's first expedition.[2]

What makes the chronicle especially intriguing is that there are few known lightning fires, and those in recent years. The resident indigenes, the Kenaitze Athabascans, have no oral tradition of lightning, much less of lightning fires. Rain is common, thunderstorms almost unknown. Instead, the Kenai's fire history over the past few centuries is one of anthropogenic fire meeting what increasingly has been a directly or indirectly anthropogenically influenced landscape. That makes the Kenai different from the interior, but perhaps not so different from the interior's future.

On his third voyage, in October 1778, Captain James Cook sailed past Kenai into what became Cook Inlet. He spotted smoke, which (as was typical) he interpreted as a sign of human inhabitants. There were landscape fires, likely from Native sources, 70 years earlier in 1708, and later in 1862. Cook's reports on sea otters—one of the few items that China would trade for—set off a fur rush. Russian *promysleniki* (trappers) from the eastern movement of the fur trade across Siberia followed. In 1787 the Russian American Company established a post on Kenai, Fort Nikolaevskaia (near the mouth of the Kenai River). Prolonged contact with the Natives usually led to a demographic decline precipitated by disease, coercion, and social breakdown. Certainly this is what happened in the Aleutians and along the Pacific Northwest; the particulars on the Kenai are unknown. How this might have translated into fire is murky. Peoples dependent on caribou tend not to burn deliberately, except in small patches or along traplines, because lichen is both the primary surface fuel and the prime winter forage, and it takes decades to recover from a fire.

The chronicle of known fires beat on—1801, 1828, 1833, 1834, 1849, 1867, 1874, 1888, and 1898, and from unknown sources in 1871, 1883, 1891, and 1910. The arrival of Europeans and Americans coincided with the departure of the Little Ice Age. The discovery of gold sparked a rush in the 1880s, which continued into the 1890s, probably leading to as many sparks as nuggets. That was followed by coal discoveries. Roads and railways sought to run from Seward to Anchorage—more slash, more sparks, more fires, though most were confined to the broad corridor defined by those rights-of-way. The railroad became a particular source of irritation, indifferent as it was to cleaning up slash or preventing fires (it was hemorrhaging money).

The landscape adjusted. Wolves vanished. Eagles had a bounty placed on them. Salmon nearly went extinct locally. Caribou disappeared between 1906 and 1917. (Visiting in 1952, Starker Leopold and Fraser Darling attributed the extinction to widespread fires that had wiped away the arboreal lichens that supplied the herds' winter range, the same lichens that powered most spruce burns.) The fires that likely drove the caribou out also enticed moose in; in general, more burns, if the patches are not too widespread, means more moose. Moose apparently expanded into the big-mammal vacuum, finding fresh fodder in the new postburn landscapes, before nearly collapsing under the onslaught of hunting, much of it for trophies. Requests for game protection led to the Kenai National Moose Refuge, established in 1941, nine days after Pearl Harbor. Following ANILCA the mission of the refuge was broadened into the maintenance of general habitat for many species, though moose have remained a charismatic core.[3]

The modern fire history of the refuge began when landscape fire met industrial fire.

In 1947 road crews constructing the Sterling Highway started the Skilak fire that swept over 300,000 acres. As an early season burn it was not notably severe, but it remade a large swathe of the refuge. The year was, for the young Alaska Fire Control Service, a trial by fire, well beyond the Kenai, but the aftershocks stayed in Alaska. A decade later the Kenai boasted the first oil and natural gas discovery in Alaska, galvanizing a new mineral rush with its attendant roads and sparks, this time fueled by and for internal combustion. Some 1,500 miles of seismic survey lines were cut through the refuge; roads were hacked everywhere; a rectangular block east of Soldotna (six townships in all) was officially removed by Congress from the refuge and privatized. In 1969, amid a deep drought, two abandoned campfires from oil exploration crews led to serious burns in August. The Russian River fire blackened 2,570 acres and the Swanson River fire 79,000 acres. This time the fires scoured deeply into the soil and led to type conversion, and this time they rumbled through the American fire establishment.

The Swanson River fire became a national story. It was the most expensive fire to suppress in American history to that time. It coincided with the Bureau of Land Management's fast-emerging fire program and with the operational opening of the Boise Interagency Fire Center. It prompted alarms over the character of industrial fire suppression as some 50 bulldozers worked the lines (so many dozers were available because of the oil field development, which brought roads and a build out of settlements). The eroding tracks left behind scarified the soil with effects that outlasted any biotic effects of the burn; a new combustion order was literally impressing itself on the old. Suddenly, Alaska was not simply a place to which wildland fire protection could be sent. It was a place with the power to unmoor the national system. An important symposium, Fire in the Northern Environment, followed. Alaska began to look more like the rest of the country in its fire problems.[4]

However the various pieces had come together in previous times they began to disaggregate. Research hints at an inflection point around 1968 when the creeping consequences of global warming seemingly crossed a threshold on the Kenai. Between 1985 and 2000 bark beetles killed nearly a million acres of spruce, mostly white and Lutz, the longest outbreak in North American history, its epicenter at the Caribou Hills. The Kenai biota began reshuffling: white spruce and hemlock spread outward, shrubs moved up slopes, peatlands began drying, yielding to shrubs and spruce, and, thanks to beetle kill and salvage logging that opened the soil to sunlight, bluejoint grass expanded, its dense rhyzomic roots crowding out young trees. The old burns, especially the 1947 fire, had reestablished black spruce that 60 years later were primed to burn as crown fire again. People who had lived in small villages, who had once harvested fish and marine mammals, and the occasional big game, were replaced by the exurbs of Anchorage and an industrial society fueled by oil. The eastern lowlands went into the Kenai Wilderness (68 percent of the entire refuge). The western lowlands rubbed against Alaska's version of sprawl, acquiring a 175-mile border of towns, strip malls, and feral cabins strung along the Sterling Highway. (During the 2009 Caribou Hills fire, crews discovered over 200 cabins not on official registers, just tucked away in the woods.)[5]

Three trends began to harden, then to align like tumblers in a lock. One, the landscape dried, as it had since the end of the Wisconsin glacial period, quickening after the Little Ice Age, and tacking again around 1968, reconfiguring the biota into new patterns. The massive bark beetle outbreak was one likely outcome; the appearance of black spruce and shrubs on formerly sphagnum-moss peat is another (covering some 60 percent of the peat since 1950), and so is the propagation of grasses. Two, ignitions have increased. More people, more starts, deliberate or accidental—this is to be expected. The surprise, however, is the appearance of lightning fires in areas for which there is no historic record or traditional recollection. Three, the ability to respond is shrinking. Wilderness on the east prevents active measures, so does road-inspired sprawl on the west. Human settlements have thickened and splashed outward, trailing sparks like beer cans and shell casings. Roads cut for oil exploration became points of entry; unregistered cabins sprouted like morel mushrooms. Within the remainder of the refuge, treatments must not harm the habitat and species for which the refuge was established. Experiments during the 1960s in mechanically improving moose habitat proved expensive and inconclusive, and are not likely to be repeated. Still, the threats were too great and close at hand to overlook. A small fuels program began working the town edges. The erosion of budgets—the Fish and Wildlife Service, the Alaska Department of Forestry, which handles fire protection—leaves little surplus to experiment and small margin for error.[6]

The fire scene picked up in the 1990s. In 1991 the Pothole Lake fire burned 7,900 aces; in 1994, the Windy Point fire, 2,800 acres; the 2004 Glacier Creek fire, 6,900 acres. The 2005 Irish Channel fire, kindled by lightning, burned 1,100 acres of mountain hemlock—a doubly unprecedented event. The fires began to move south. They became scary in 2007 when the Caribou Hills fire, started by sparks from a cabin resident sharpening his shovel on a grindstone, burned 55,000 acres and 197 structures. In 2014 the Funny River fire burned both biotic borders, one along the outskirts of Soldotna, the other into alpine tundra. Had a shaded fuelbreak not been put in years earlier and the originating wind not blown from the north, the consensus is that the fire would have crashed into town. The next year the Card Street fire struck near the same area and burned three houses before swarming into the receptive refuge.[7]

Both the Funny River and the Card Street fires began in that anomalous rectangle that oil and gas discovery caused to be excised from the refuge. The two realms of combustion were interacting in ways that no one would have predicted. So we can add fire to the roster of ecological processes that are affecting the Kenai not so much by themselves but in unexpected synergies with the others.

The issues that plague the Kenai are those typical throughout fire-prone Alaska; the pieces just have somewhat different dimensions and combine in peculiar ways. Maybe the surest assessment of what is happening is that the Kenai is reasserting, in tongues of flame, the sense in which it is transitional, that it functions as a portal. It's the warmest and wettest subregion of the state, save the coastal southeast. Here the coming order of fire may be first entering Alaska. It's a future that appears to promise more fire with less control.

Almost uniquely over the past 40 years, Alaska has succeeded in keeping fire, good fire, on the land. But the constraints are growing to shrink the area available for ecological burning, and the fires that are coming are not, by traditional standards, unequivocally good. They are burning differently. They are catalyzing the effects of change in ways that may not be restoring, or even maintaining, existing fire regimes, so much as kindling novel ones. This may be exactly what the land needs as the wave train of changes continues—fires as a constant series of ecological jolts and twitches that lessen the grand shock of cumulative change. Or they may serve as an accelerant to those changes, shifts that people will be sorry to see.

The refuge knows it must keep its fires (and their smokes) out of the towns. It knows the public expects a moose refuge to have moose. It knows the prime movers behind change are beyond its control. But with so much in flux, it is not simple to identify future desired conditions or to specify prescriptions to achieve them. Without a clear, usable past, managers are left with a series of ad hoc adjustments, hoping that their actions and fires will ease the refuge into a usable future.

But then it's not clear what that future *should* look like. On the Kenai the past seems less like a prologue than an endless present. The future may be unsettled in ways the past doesn't foreshadow. This is not a stable place that is now changing, but a changing place that is poised to change faster. In that respect, the Kenai fire scene might well stand for most of Alaskan fire. Or for that matter, for the Alaskan Anthropocene.

NORTH TO THE FUTURE

Pleistocene to Pyrocene

I N 1898, AS KLONDIKE FEVER raged, the U.S. Geological Survey published a map of Alaska to show the known sites of mineral wealth. The most prominent locales, of course, were the gold fields. But not far behind were coalfields. Fossil fuels have been the black gold of Alaska since the beginning of American rule.

Twice those lithic landscapes have entered the national narrative of environmentalism. In 1909, after Teddy Roosevelt had left office, Gifford Pinchot, then chief forester, picked a fight with Secretary of the Interior Richard Ballinger over the leasing of Alaskan coal lands. The particulars of the leasing were murky at best, but the coal was only a means (to Pinchot's mind) to address larger questions of how to manage the nation's estate after Roosevelt's departure. The resulting controversy forced President William Howard Taft to fire Pinchot, precipitating a major split in the Progressive movement and the Republican Party, and leaving a bitter legacy for conservation. At the time Alaska was two years away from status as a bona fide territory.

Sixty years later it was Alaskan oil from the North Slope that ignited a national uproar. The oil could only come to market by constructing a pipeline from Prudhoe Bay to Valdez. Alaska had been a state for 10 years and suffered from a feeble economy in need of defibrillation. The controversy proceeded along classic Alaskan lines between those who wanted the Wild and those who wanted a Wild West. Those who didn't want the Arctic drilled for oil, those who worried about oil spills from a metal tube traversing cold, seismically active terrane, those who feared ecological disruption, especially for migratory wildlife and from roads branching into every nook and cranny—all wanted the pipeline stopped. Those who wished for Alaska to develop along the lines of the Lower 48, who desired a chance to advance a new frontier, who sought a state budget that could provide for basic services—all wanted the pipeline built. The process stalled until 1971 when the Alaska Native Claims Settlement Act resolved the issue of land tenure, and it found a greased legislative rail with the first OPEC oil embargo.

The energy crisis ensured the pipeline would be built. The long controversy, however, meant it would be designed to withstand the known hazards of the Alaskan landscape and the threats a pipeline might pose. It was constructed to withstand earthquakes, permafrost, and intense cold. It allowed the movement of fauna and the migration of caribou. And it was hardened against the heat posed by a boreal forest fire. Along its wending way the two grand realms of combustion meet.

The pipeline divides the state geographically and historically, and during its construction, it divided it politically. It provoked major controversies that have defined the economic geography and political economy of modern Alaska. It laid down the basis for Alaska as a modern petro-state. The oil industry accounts for a third of the state's economy; and between 80–90 percent of state revenue comes from oil taxes, rents, and royalties. This much is widely understood. Less appreciated is that the pipeline symbolized, and makes possible, a combustion divide. It segregates two eras of Alaskan fire history.[1]

On most popular maps, and in the public imagination, two great paths cross Alaska. One reenacts the past, one leads to the future. The Itidarod trail runs from Willow (near Anchorage) to Nome, and it's the site for an annual dogsled race that harkens back to Alaska's first mineral rush. The other wends from Prudhoe Bay to Valdez, and through it flows the oil that lubricates the modern Alaskan economy.

Those two routes can also stand for two paths of Alaskan fire. The Klondike Gold Rush, which eventually spilled over most of the state, was rife with fires. Fires cleared the woods to expose outcrops, fires melted permafrost to reach placers; fire was life, for without it there was no resistance to the killing cold. The pipeline speaks to another kind of fire, without which modern Alaska could not exist. Fossil fuels heat homes, run power stations, fuel vehicles, and fill (or not) the coffers of the state legislature. Thanks to oil, Alaska has no income tax, no sales tax, and in most years sends a rebate to citizens. If the flow of those fuels falters, the state suffers. If something extinguishes the fires it feeds, the state would collapse.

Both fires—those that burn living landscapes and those that burn lithic ones—continue; and both are projected to increase in coming years. The linkage between them is worth exploring. There are instances of direct competition, where flames burn along the pipeline, and where smoke from wildland fires has forced the turbines running pumps to shut down. Mostly, however, the interaction is indirect.

The society that oversees wildland fire management is a profoundly fossil fuel–based civilization. Industrial combustion supplies the vast bulk of its energy needs. That society runs on machines, literal fire engines, that burn fossil fuels. It no longer exists on a subsistence level because its industrial fire economy can supplement what it can produce locally by importing goods and services from around the globe. Industrial combustion makes possible its wildland fire program. Fire management operations run on trucks, engines, pumps, helicopters, aircraft, even driptorches—all burning gas—that get firefighters or water to the fireline. Fire management policy is made in offices lit by electricity, heated and cooled by electricity or gas, and over desks with telephones and computers powered by off-site dynamos burning coal or oil. State firefighters are paid with the revenue derived from oil and coal leases. Emergency firefighter crews choose between a traditional economy based on open flame and one that houses fire in machines.

But there is a still deeper interaction, which lies in fire's capacity to mobilize carbon. Greenhouse gases liberated by burning taiga and tundra are joining those far more pervasive gases spewing out tailpipes and smokestacks, and together they are unmooring the climate that human society has adapted to over the past 6,000 or so years. Projections suggest that the atmospheric warming subsequently created will alter—must alter—the existing arrangement of fire regimes. Most models posit an increase in factors that will dry fuel; many also suggest an uptick in lightning. The nightmare scenario holds that the resulting big burns will set up a positive feedback loop such that the large carbon-storing woods will convert to little grasses and shrubs, freeing thus more carbon. The two grand realms of combustion will no longer compete so much as collude. The Anthropocene will become a Pyrocene.

―――――――

Yet once again, Alaska is different.

Much as its Natives upset the traditional dialectic between economy and ecology, so its immense tundra is destabilizing the traditional discourse about the two prevailing forms of combustion. The third party here is organic soil. Those soils constitute huge carbon sinks, whether as boreal peat or embedded in permafrost. They unbalance the usual calculations of competition and collusion.

There are, in fact, three fuels in play. One lies above the surface—the woods, shrubs, grasses, mosses, lichens. One lies far below the surface in the form of ancient biomass, sequestered in the sediments of deep time. And one is the shallow subsurface biomass still lingering from the Pleistocene, whose planetary frost and thaw cycles left big reservoirs in the not-too-distant ground. If liberated by fires, or climatic warming, or an intricate choreography of warming and burning, they could push the planet quickly into a tipping point from which a return might be impossible. Those frozen reserves are so immense that they transform a combustion dialectic into a braided narrative. How these three fires amplify, dampen, and leverage one another is the evolving story of fire in Alaska.

It is unclear whether fires in tundra are increasing. Soil charcoal hints that fires have occurred in the past, though nothing, over the past 5,000–7,000 years, on the scale of the 2007 Anaktuvuk burn on the North Slope. It's too early to know whether such outbreaks are the harbingers of a new regime or simply reflect the short memory and lean data sets of recording Alaskans. The calculations of what quantities of carbon might be released if the burning infects the land into a combustion contagion are terrifying. What is clear is that the Alaskan fire scene may change in ways that escape the grasp of our present anecdotes and algorithms.

It may be that the permafrost is another Pleistocene relic that has lingered for millennia, surviving when wooly mammoths and giant ground sloths and other emblems have vanished. One after another, the relics of the Ice Age are going. The glaciers are receding, the arctic pack ice is shrinking, the annual snowpack is smaller. Permafrost survived because it was insulated. But if fires and warming peel back that layer of organic insulation, it will go the way of Pleistocene ice generally. A Fire Age will have driven off the last vestiges of the Ice Age.

Alaska may be America's great refuge for ice, as it has been for wolves and brown bears. What should not be forgotten is that the ice has been a check on fire, and fire on ice. This is not so much a planetary dialectic as a dialogue. In the 1970s climate scientists warned of a coming ice age. The Milankovitch cycles still spin, the Earth's oceans and continents are still aligned favorably to leverage snow into ice sheets, and 80 percent of the past 2.6 million years has been glacial—there is no reason to think that our brief interglacial will persist for much longer. It had already lasted longer than models predicted. The ice was coming. The Little Ice Age was a warning shot off the bow that a big ice age was inevitable.[2]

It didn't happen, and now it seems it can't. We've halted the ice. But in stopping ice we've unleashed fire. We're a fire creature: we can exist without ice but not without fire. But we may be knocking away all the constraints that have traditionally kept fire within the bounds of usefulness. The Alaskan fire scene may be where we drive the last of the Pleistocene ice into extinction, or whether, alternatively, we recognize that we are also a creature of the Pleistocene and may be unwisely turning our firepower against ourselves.[3]

Like the rest of America, Alaska seems caught between two fires. Unlike the rest of the country, it still has room, within limits, to trade space for time. Those choices will undoubtedly involve national as well as Alaskan politics, as they should, since those choices will affect us all.

EPILOGUE

Alaska Between Two Fires

IN 2009 THE RUTH FIRE started in Denali National Park. What made it distinctive was that it burned atop the Ruth rock glacier, which had acquired a combustible crust of vegetation. The fire was allowed to burn itself out.

At first glance the episode might be synecdoche for modern wildland fire management in Alaska: a nonthreatening fire was allowed to free-burn away. In the (not-so) old days, the Ruth fire would have been fought. Now, like wolves, it was permitted to live out its natural life. A deeper peering, however, might note that the real curiosity in the scene is the overlay of fire on ice. The Ruth fire testified not just to the modern evolution of fire policy but to the *longue durée* of Alaskan fire history. The coming age is likely one in which the ice will recede and the fires flourish.

In one of the great pieces of Alaskan literature, "To Build a Fire," Jack London depicts a struggle against the cold. The everyman protagonist's survival depends, finally, on his ability to light a warming fire. He finally succeeds, but kindles those fledgling flames beneath a tree that drops snow on them and extinguishes his last hope. Today's version would invert that relationship. It is the fires that are extinguishing the ice. London's human needed fire to keep the ice at bay. Now, we need to manage fire to keep the ice. Our survival depends not on lighting more and bigger fires but on keeping them within bounds. Our massing firepower has a reach beyond our grasp.

It's not easy to place Alaska in the American fire scene. Few maps (certainly none from the Lower 48) show Alaska in its proper geographic setting; many kids probably know it as a state the size of South Dakota sited off the coast of Baja California. It's similarly hard to place Alaska into the national fire narrative because, again, it sits not just to the margins but outside the frame of the chronicle. Almost all of the national story can be written with no more reference to Alaska than to Tennessee or Nebraska. New York, Oregon, Michigan—all have influenced national policy more than Alaska. To place it geographically, you have to widen the aperture to include North America. To locate it historically, you have to reach for other narratives than the Received Standard Version.

Yet Alaska matters.

It showed how, within a handful of years, to scale up a fire suppression program from nothing to a major presence, and then how to transfer that program to another part of the country as the BLM did between Alaska and the Great Basin. It showed how to base fire protection on aircraft. It showed how to create a fire culture outside the Forest Service. It showed how to use remote sensing to assist fire management. After the fire revolution, it showed how to restore fire on a landscape scale—one of the few places that succeeded, particularly with natural fire, and far and away the most expansive expression. Florida created a model based on prescribed fire. Alaska demonstrated one based on managed wildfire. It showed how to integrate a single fire-suppression service, needed for efficiency, with various fire management policies, needed to be effective. There are places in the Lower 48 that have joint operations among the Forest Service, Bureau of Land Management, Fish and Wildlife Service, and National Park Service, and there are fire monitoring and prescribed fire modules that serve many units and agencies, but there are none that operate on the scale characteristic of Alaska, and few that involve states as truly equal partners. And it has shown with unblinking clarity the two competing realms of combustion—one that burns living landscapes and one that burns lithic landscapes—that define not only America but the Earth.

In exploring the character of Alaskan fire as a "wicked problem," Stuart Chapin and his colleagues note the paradox, which becomes a dilemma, of Native villages. They have traded the mobility of a subsistence economy for fixed residences and some modern amenities, yet they value and wish to preserve traditional skills and culture. Human societies and free-burning landscape fire, however, coexist only if the people move—if they syncopate their practices with the successional phases of postburn landscapes. Some parts burn on rhythms of 1–10 years, some on 10–30 years, some on 50–60 or longer. By adopting fixed residences, the villagers lose that ancient relationship to fire. To maintain their new lives, they want fires suppressed so that the trapping, hunting, and foraging can continue nearby, and they are happy with fire-suppression jobs that bring in needed cash. In the short term all this makes sense. In the long term it could mean the deterioration, possibly the collapse, of the sustaining biota. Nature's economy needs its specie in circulation, not buried in caches or stuffed in mattresses. The long-term consequence means the replacement of patchy burns, scattered over space and time, with conflagrations, some of which will likely threaten the villages themselves.[1]

Yet this is the same dilemma that all of Alaska faces. As a petro-state, its economy depends on fossil fuels, mostly oil, but with coal waiting in the margins. Even Native villages run on ATVs, powerboats, chainsaws, pumps, snowmobiles, and electricity, and need to earn cash to pay for those goods. Urban Alaskans, too, seem to want to preserve a culture they identify with pioneering, or sourdoughs, or at least the autonomous freedoms those worlds promise in the popular mind. They are less excited about untrammeled wilderness than untrammeled agency. But the fire economy that sustains their current lives threatens to unhinge the climate upon which the monumental, often mythical Alaska they appeal to depends. The high latitudes will feel the effects of global change far more than temperate ones. The short-term solution is to pump everything dry, which allows urban Alaskans (which is to say, 70–80 percent of all Alaskans) to live as modern Americans. The long-term consequence is likely to make Alaska a very different place, and not in ways that Alaskans say they like.

Two different economies based on combustion, two parallel choices to change the way a society lives with fire. The upshot in both instances is likely to be unhappy, perhaps ruinous. Only some outside force can possibly intervene to alter the outcome. That means government, and the experience and antistate ideology of neither party wants that. It's not just a wicked problem: it's a wick problem. It involves choices among combustions.

Maps of ignition by cause track two distinct realms of Alaskan fire. They seem to interact only along the margins where a fossil-fuel civilization meets free-burning landscape flames. The reality is different. The smoke of burning landscapes enters the nominal fire protectorates around urban sites. Fire-powered engines allow landscape fires to be managed, even extinguished. Greenhouse gases destabilize the climate that all Alaska shares. The two realms of combustion interact in ways that are poorly understood. Like an oil drop suspended between charged plates, Alaska lies between those two kinds of fire.

Today, combustion is both an enabler and a presence. What mix of burning Alaskans choose will decide how they balance economy and ecology, desires and fears. It will decide whether the Last Frontier, however that gets interpreted, will instead devolve into a Lost Frontier.

———

Alaska so big, so looming, so seemingly immune to human meddling, still so much Alyeska, the Great Land, as the Aleuts called it, that it can't help but impress itself on any one who stares across its distances. It seems impossible that humans could do more than scratch its surface or whistle in its williwaws.

But repeatedly, people have done just that. They have in the past nearly exhausted its fisheries, its fur seals, its sea otters; they have slashed and burned many of its forests; they have disrupted, perhaps pushing toward local extinctions, its big game; they have overturned large swathes of its soils; they have unsettled its fire regimes; and by combusting its lithic landscapes they promise to push its post–Ice Age climate into a Fire Age that will melt its mountain glaciers and permafrost plateaus. They are replacing the biotic and geomorphic relics of the Pleistocene with industrial surrogates of the Anthropocene. These are not trivial effects. Alaska is big, but it is not bigger than the Anthropocene.

What has kept these events from being fatal is Alaska's awkward relationship to the United States. The nation has both stifled and saved Alaska repeatedly. Reserves and restrictions imposed by a force larger than colony or state or big business have kept Alaska from sinking into the social equivalent of muskeg. Compared with western states, its Act of Statehood is remarkably generous and progressive. Yet it's an uneasy bonding, at times an unhappy codependency, and it strikes to the core of the Alaskan persuasion.

Like all matters Alaskan that persuasion ultimately goes back to land. It's what people have fought over—all the great controversies hinge on control over land, or its future promise. It's also what allows for the fights to occur. Alaska's natural estate still has room for maneuvering, space for experiments and new thoughts, a horizon not crowded beyond human sight. Its immense land base has allowed for the state, Natives, and nation to have a say and oversight over what most animates them. It's behind the innovations in Alaskan fire. It remains the premise of Alaska's future.

Alaska is the place where anthropogenic fire first entered the New World. It may be the place where the Anthropocene, powered by industrial combustion, first manifests itself in the Americas.

NOTE ON SOURCES

A LASKA HAS BEEN a fun region to research. Mostly, of course, I have to hew to the scientific and technical literature on fire, which is largely published in journal articles. I found a good synthesis, however, in F. Stuart Chapin III et al., *Alaska's Changing Boreal Forest*.

There are wonderful background books for Alaskan history. I began with Ernest Gruening, *The State of Alaska*, which I first read in graduate school. Two academic historians have written widely about Alaska: Morgan Sherwood, with *The Exploration of Alaska, 1865–1900* and *Big Game in Alaska: A History of Wildlife and People*; and Stephen W. Haycox, *Battleground Alaska: Fighting Federal Power in America's Last Wilderness*, among many other books. I also found useful as an introduction Roger W. Pearson and Marjorie Hermans, eds., *Alaska in Maps: A Thematic Atlas*. And I'd be remiss not to mention Peter Coates, *The Trans-Alaska Pipeline Controversy*, which ranges far beyond its nominal title. For institutional histories prior to 1980, I relied on my previously published account in *Fire in America*. Mike Roos has gathered a helpful collection of the gray literature on the creation of the Alaska Fire Service, which he generously made available to me.

Otherwise I talked to fire folks—always informative, always a pleasure.

NOTES

AUTHOR'S NOTE: OAK WOODLANDS

1. From Appendix A, "The Name 'Kentucky,'" in Robert F. Collins, *A History of the Daniel Boone National Forest 1770–1790*, ed. Betty B. Ellison (U.S. Forest Service, Southern Region, 1975).

PROLOGUE: EAST OF THE 100TH MERIDIAN

1. For a fuller rendition, see Frederick Jackson Turner, *The Significance of the Frontier in American History* (New York: Henry Holt, 1921), 12.

THE LONG HUNT

1. John Mack Faragher, *Daniel Boone* (New York: Henry Holt, 1992), 43–44.
2. For a brief survey of human settlement, see A. Gwynn Henderson, "Dispelling the Myth: Seventeenth- and Eighteenth-Century Indian Life in Kentucky," *Register of the Kentucky Historical Society* 90, no. 1 (1992): 1–25. Henderson makes the point that "Kentucky" had been occupied for a long time by settled peoples. But the Barrens and Bluegrass do seem to have been used as seasonal hunting grounds, which is my interest. At the time of colonial contact, those lands do appear as seasonally vacant, as reported. My interest of course lies in their fire history. They were maintained by regular burning. Also recommended for its lively synthesis of the existing literature is Andrew P. Patrick, "Birth of the Bluegrass: Ecological Transformations

in Central Kentucky to 1810," *Register of the Kentucky Historical Society* 115, no. 2 (Spring 2017): 155–82.

3. Robert F. Collins, appendix A of *History of Daniel Boone National Forest, 1770–1970,* ed. Betty B. Ellison (U.S. Department of Agriculture, Forest Service, Southern Region, 1975), https://foresthistory.org/wp-content/uploads/2017/01/A-history-of-the-Daniel-Boone-National-Forest.pdf.

4. The folk burning was not restricted to the Barrens: it accompanied the settlers across the Appalachians. For a useful survey, see William H. Martin, "The Role and History of Fire in the Daniel Boone National Forest" (Daniel Boone National Forest, n.d.).

5. Carl O. Sauer, *Geography of the Pennyroyal.* Kentucky Geological Survey 6, vol. 25, 1927, 84, 143. Donald Davidson et al., *I'll Take My Stand: The South and the Agrarian Tradition* (New York: Harper, 1930; Baton Rouge: Louisiana State University Press, 2006).

6. The best summary remains Paul A. Delcourt et al., "Holocene Ethnobotanical and Paleoecological Record of Human Impact on Vegetation in the Little Tennessee River Valley, Tennessee," *Quaternary Research* 25 (1986): 330–49, followed by Paul A. Delcourt et al., "Prehistoric Human Use of Fire, the Eastern Agricultural Complex, and Appalachian Oak-Chestnut Forests: Paleoecology of Cliff Palace Pond, Kentucky," *American Antiquity* 63, no. 2 (1998): 263–78; for a more distilled version see Hazel R. Delcourt and Paul A. Delcourt, "Pre-Columbian Native American Use of Fire on Southern Appalachian Landscapes," *Conservation Biology* 11, no. 4 (August 1997): 1010–14.

7. For an excellent primer, see Patrick H. Brose, Daniel C. Dey, and Thomas A. Waldrop, *The Fire-Oak Literature of Eastern North America: Synthesis and Guidelines,* General Technical Report NRS-135, U.S. Forest Service, 2014.

8. Kenneth B. Pomeroy, "Fire Conference," *American Forests* 66, no. 1 (January 1960): 24.

A DARK AND BURNING GROUND

1. I follow Carl Ortwin Sauer, *Geography of the Pennyroyal: A Study of the Influence of Geology and Physiography upon Industry, Commerce and Life of the People,* Kentucky Geological Survey, 6, vol. 25, 1927, 123–30, reproduced in John Leighly, ed., "The Barrens of Kentucky," in *Land and Life: A Selection from the Writings of Carl Ortwin Sauer* (Berkeley: University of California Press, 1963).

2. François André Michaux, *Travels to the West of the Allegheny Mountains* (1895), reprinted as vol. 3 of Reuben Gold Thwaites, ed., *Early Western Travels, 1748–1847* (Cleveland, Ohio: A. H. Clark, 1904), 221–22.

3. Michaux, 268.

4. R. W. Wells, "On the Origin of Prairies," *American Journal of Science and the Arts* 1, no. 4 (1819): 335.

5. Wells, 335.

6. Jefferson to Adams, May 27, 1813, quoted in "Thomas Jefferson on Forest Fires," *Fire Control Notes* 13 (April 1952): 31.

7. All quoted in Sauer, *Geography of the Pennyroyal*, 123–30.

8. Sauer, 30. For a lively distillation of forestry law and institutions in 19th-century Kentucky and Tennessee, see Ralph R. Widener, ed., *Forests and Forestry in the American States: A Reference Anthology* (National Association of State Foresters, n.d.), 322–37, 308–12.

9. Betty Joe Wallace, *Between the Rivers: History of the Land Between the Lakes* (Clarksville, Tenn.: Austin Peay State University, 1992), 172.

10. Julian A. Steyermark, *Vegetation History of the Ozark Forest*, University of Missouri Studies (Columbia: University of Missouri Studies, 1959), 51.

11. Steyermark, 50–53.

12. See Michael Williams, *To Pass on a Good Earth: The Life and Work of Carl O. Sauer* (Charlottesville: University of Virginia Press, 2014).

13. Sauer, *Geography of the Pennyroyal*, 128; Carl O. Sauer, "The Agency of Man on Earth," in *Man's Role in Changing the Face of the Earth*, vol. 1, ed. William L. Thomas Jr. (Chicago: University of Chicago Press, 1956), 55. See also Sauer, "Fire and Early Man," in Leighly, *Land and Life*, 288–99.

14. Sauer, "The Agency of Man on Earth," 54–56. James J. Parsons, "Obituary: Carl Ortwin Sauer, 1889–1975," *Geographical Review* 66, no. 1 (January 1976): 86. Sauer, "Fire and Early Man," 297–98.

15. See Carl Sauer, "Man's Influence upon the Earth," *Geographical Review* 1 (1916): 462.

16. Marc C. Abrams, "Fire and the Development of Oak Forests," *BioScience* 42, no. 5 (1992): 346–53.

17. Mary A. Arthur et al., "Refining the Oak-Fire Hypothesis for Management of Oak-Dominated Forests of the Eastern United States," *Journal of Forestry*, July/August 2012, 257–66. A fuller, later version is available in Patrick H. Brose, Daniel C. Dey, and Thomas A. Waldrop, *The Oak-Fire Literature of Eastern North America: Synthesis and Guidelines*, General Technical Report NRS-135, U.S. Forest Service, 2014.

UNCHANGED PAST: STONES RIVER NATIONAL BATTLEFIELD

1. I wish to thank Jesse Burton, Travis Neppl, and Gilbert Backlund for organizing an informative tour of Stones River and introducing me to the complexities of the current battleground over fire and ecological health.

UNCERTAIN FUTURE: LAND BETWEEN THE LAKES

1. I wish to thank Dennis Wilson for a vigorous field introduction to LBL at a time when his calendar was crowded and the calls on his attention many.

2. My primary sources are Wallace, *Between the Rivers*, and Edward W. Chester and James S. Fralish, eds., *Land Between the Lakes, Kentucky and Tennessee: Four Decades of Tennessee Valley Authority Stewardship* (Clarksville, Tenn.: Center for Field Biology, Austin Peay State University, 2002). Sauer quote from *Geography of the Pennyroyal*, 88.

3. A good chronology exists in Thomas D. Forsythe, "The 'Land Between the Lakes Area Biosphere Reserve'—Can It Be a Global Model for Sustainable Development?," in Chester and Fralish, *Land Between the Lakes*, 169–81.

4. Ronald A. Foresta, *The Land Between the Lakes: A Geography of the Forgotten Future* (Knoxville: University of Tennessee Press, 2013), 2–3.

UNSETTLED PRESENT: NATURE CONSERVATION

1. I wish to thank Shelly Morris, Chris Minor, Jeffrey Sole, and the rest of the TNC staff for a wonderful introduction to Mantle Rock and its fire program.

2. Albert J. Meier and Todd Jobe, *Fire and Disturbance History of Mantle Rock Preserve*, final report submitted to the Nature Conservancy, December 28, 1999.

3. Mike Stambaugh, "Wave of Fire," in press. A videotaped lecture of the central argument is available at https://mediasite.video.ufl.edu/Mediasite/Play/48f73e5279f24997b4bc0150126ac75b1d.

4. I wish to thank Joyce Bender for a marvelous primer on Kentucky state nature preserves and the fire ecology of its relic glades and barrens. Also, Elizabeth Wright for her queries to Joyce, which enlarged my meager background in botany.

5. While I focus on the Shawnee, my field tour ranged widely through the institutions of southern Illinois. The following people made that tutorial

possible: David Allen, Scott Crist, Dave Jones, Jesse Riechman, Jody Shimp, Benjamin Snyder, and Charles Ruffner, who indefatigably organized the program.

6. John L. Nelson et al., "Drainage and Agriculture Impacts on Fire Frequency in a Southern Illinois Forest Bottomlands," *Canadian Journal of Forest Research* 38, no. 12 (2008): 2932–41.

7. William L. Thomas Jr., ed., *Man's Role in Changing the Face of the Earth*, 2 vols. (University of Chicago Press, 1956); David Potter, *People of Plenty: Economic Abundance and the American Character* (Chicago: University of Chicago Press, 1954).

MISSOURI COMPROMISE

1. This chapter is a species of interpretive journalism that resulted from a two-day field trip to the Missouri Ozarks organized by Rich Guyette, Dan Dey, and Mike Stambaugh, as a prelude for a daylong workshop on human fire history at UM–Columbia. For some years I have followed the fascinating fire-history articles the UM Tree-Ring Lab group had published and leaped at the chance to see them and their sites in person. Others joined in: Tim Nigh, Susan Flader, Dan Drees, and Rose-Marie Muzika. To their research I have tried to provide a larger historic and philosophical context. The data is theirs. The refractive prism is mine. I also thank Mike Dubrasich for a gentle editing of a rough-pixelated manuscript.

2. Milton D. Rafferty, *Rude Pursuits and Rugged Peaks: Schoolcraft's Ozark Journal 1818–1819* (Fayetteville: University of Arkansas Press, 1996), 62–63.

3. The classic introduction remains Carl O. Sauer, *The Geography of the Ozark Highland of Missouri* (New York: Greenwood Press, 1968).

4. See, for example, Curtis Marbut, "The whole region and its vegetation was more closely allied to the western prairies than to the timber-covered Appalachians." Quoted in Tim A. Nigh, "Missouri's Forest Resources—An Ecological Perspective," in *Toward Sustainability for Missouri Forests: Proceedings of a Conference*, ed. Susan L. Flader, General Technical Report NC-239, U.S. Forest Service, 1999.

5. See Michael J. O'Brien and W. Raymond Wood, *The Prehistory of Missouri* (Columbia: University of Missouri Press, 1998), 295–96, 331–33.

6. Sauer, *Geography of the Ozark Highland*, 52–54. Quotes on Indian burning from Marbut come from Nigh, "Missouri's Forest Resources," 11.

7. Leopold quote from Susan L. Flader, "History of Missouri Forests and Forest Conservation," in Flader, *Toward Sustainability for Missouri Forests*, 20.

8. Sauer, *Geography of the Ozark Highland*, 207, 230–33, 237.

9. See E. R. McMurry et al., "Initial Effects of Prescribed Burning and Thinning on Plant Communities in the Southeast Missouri Ozarks," *Proceedings of the 15th Central Hardwood Forest Conference*, U.S. Forest Service e-GTR-SRS-101 (2006): 241. The most comprehensive summary of contemporary fire statistics is Steve Westin, "Wildfire in Missouri" (Jefferson City: Missouri Department of Conservation, 1992).

10 Details of conservation history from Flader, "History of Missouri Forests and Forest Conservation."

11. See Susan Flader, "Missouri's Pioneer in Sustainable Forestry," *Forest History Today*, Spring/Fall 2004, 2–15.

12. The UM–Columbia group under Richard Guyette has produced an ever-lengthening literature on these topics. Perhaps the central paper is R. P. Guyette, R. M. Muzika, and D. C. Dey, "Dynamics of an Anthropogenic Fire Regime," *Ecosystems* 5, 2000, 472–86. I take considerable liberties in extrapolating their concepts into a more general critique of fire scholarship.

AUTHOR'S NOTE: PACIFIC NORTHWEST

1. Miles Wilson, "Slash Burning," in *Harm* (Reno: University of Nevada Press, 2003): 75–76.

PROLOGUE: GREEN ON BLACK

1. James K. Agee, *Fire Ecology of Pacific Northwest Forests* (Covelo, Calif.: Island Press, 1993), 8. Muir quoted in Stephen Pyne, *Fire in America: A Cultural History of Wildland and Rural Fire* (Princeton, N.J.: Princeton University Press, 1982), 327n9.

FIRE AND AXE: THE FIRST AND SECOND TIMBER WARS

1. Gifford Pinchot, *The Fight for Conservation* (New York: Doubleday, 1910), 15.

2. On the early alliance between industry and the state, see George T. Morgan, "The Fight Against Fire: The Development of Cooperative Forestry in the Pacific Northwest, 1900–1950" (PhD diss., University of Oregon, 1964).

3. For a summary, see Pyne, *Fire in America*, 338.

4. Logan A. Norris, "An Overview and Synthesis of Knowledge Concerning Natural and Prescribed Fire in Pacific Northwest Forests," in *Natural and Prescribed Fire in Pacific Northwest Forests*, John D. Walstad, Steven R.

Radosevich, and David V. Sandberg, eds. (Corvallis: Oregon State University Press, 1990): 7. Data taken from J. K. Agee, "The Historical Role of Fire in Pacific Northwest Forests," same volume, 37.

5. A nice summary of the fire is available in *The Oregon Encyclopedia*, "Biscuit Fire of 2002," at https://oregonencyclopedia.org/articles/biscuit_fire _of_2002/. The best review of the controversies is the GAO, *Biscuit Fire: Analysis of Fire Response, Resource Availability, and Personnel Certification Standards*, GAO-04-426, April 2004.

6. As of this writing, an assessment of the Chetco Bar fire is still underway. For the basics see Inciweb (https://inciweb.nwcg.gov/incident/5385/) and the Chetco Bar timeline published by the Forest Service (https://usfs.maps.arcgis .com/apps/Cascade/index.html?appid=809cc1882e8d45169b9baf2669f95c5a).

GRACE UNDER FIRE: THE WILLAMETTE VALLEY

1. Many people donated time to help me understand the historic and present fire scene in the Willamette Valley. At Willow Creek Preserve: Amanda Stamper, Ed Alverson, and Jess Gillimore. At Grand Ronde: David Harrelson, Briece Edwards, Colby Drake, and Joyce Lecomte. My thanks to them all for taking the time to explain what they know so well to someone who understood so little of it when he arrived.

2. Quotes from Robert Boyd, "Strategies of Indian Burning in the Willamette Valley," in *Indians, Fire, and the Land in the Pacific Northwest*, ed. Robert Boyd (Corvallis: Oregon State University Press, 1999), 101, 108.

3. The two classic composite summaries are Boyd, "Strategies," and Carl Johannessen et al., "The Vegetation of the Willamette Valley," *Annals of the Association of American Geographers* 61, no. 2 (1971): 286–306. A masterful survey of the landscape during early settlement is John A. Christy and Edward R. Alverson, "Historical Vegetation of the Willamette Valley, Oregon, Circa 1850," *Northwest Science* 85, no. 2 (May 2011): 93–107. A longer-term view is available in Megan K. Walsh, Cathy Whitlock, and Patrick J. Bartlein, "1200 Years of Fire and Vegetation History in the Willamette Valley, Oregon and Washington, Reconstructed Using High-Resolution Macroscopic Charcoal and Pollen Analysis," *Palaeogeography, Palaeoclimatology, Palaeoecology* 297 (2010): 273–89. For a very useful compendium of fire and restoration essays, see the special edition of *Northwest Science* 85, no. 2 (May 2011).

4. See Bob Zybach, "The Great Fires: Indian Burning and Catastrophic Forest Fire Patterns of the Oregon Coast Range, 1491–1951," PhD diss. (Oregon State University, 2003).

5. Boyd, "Strategies," 107.
6. For more background, see Christina Kakoyannis, "Learning to Address Complexity in Natural Resource Management," PhD diss. (Oregon State University, 2005), and Christopher Duerksen and Cara Snyder, *Nature-Friendly Communities: Habitat Protection and Land Use* (Washington, D.C.: Island Press, 2005), chapt. 6, "Eugene, Oregon: Shining Star of Wetlands Preservation." I'm indebted to Jessica Gallimore for the references.
7. A dutiful but somewhat drab summary of the challenges is available in Sarah T. Hamman et al., "Fire as a Restoration Tool in Pacific Northwest Prairies and Oak Woodlands: Challenges, Successes, and Future Directions," *Northwest Science* 85, no. 2 (May 2011): 317–28. For a digest of recent smoke legislation, see http://www.oregonlive.com/politics/index.ssf/2009/06/oregon_legislature_bans_field.html.

CROSSING THE KLAMATH

1. My basic reference is James K. Agee, *Steward's Fork: A Sustainable Future for the Klamath Mountains* (Berkeley: University of California Press, 2007), supplemented by Carl N. Skinner, Alan H. Taylor, and James K. Agee, "Klamath Mountains Bioregion," in *Fire in California's Ecosystems*, ed. Neil G. Sugihara et al. (Berkeley: University of California Press, 2006), 170–94. The complexity of the Klamath's conifer assemblage comes from 19; on 31 Agee says the complexity is the greatest on Earth. Alan Taylor and Carl Skinner agree; see "Fire Regimes and Management of Old-Growth Douglas-Fir Forest in the Klamath Mountains of Northwestern California," in *Proceedings—Fire Effects on Rare and Endangered Species and Habitats Conference* (International Association of Wildland Fire, 1997), 203.
2. I would like to thank Rick Young and Eamon Engber for a lovely tutorial capped by a field trip through the major habitats of the park, and for arranging a glorious day to view it all.
3. Studies on redwood fire history include Peter M. Brown and William T. Baxter, "Fire History in Coast Redwood Forests of the Mendocino Coast, California," *Northwest Science* 77, no. 2 (2003): 147–58; B. S. Ramage, K. L. O'Hara, and B. T. Caldwell, "The Role of Fire in the Competitive Dynamics of Coast Redwood Forests," *Ecosphere* 1, no. 6, December 2010, article 20; Peter M. Brown, "What Was the Role of Fire in Coast Redwood Forests?," in *Proceedings of the Redwood Region Forest Science Symposium: What Does the Future Hold?*, ed. Richard B. Standiford et al., General Technical Report PSW-GTR-194, U.S. Forest Service, 2007, 215–18; Peter M. Brown and Thomas W. Swetnam, "A Cross-Dated Fire History From Coast

Redwood Near Redwood National Park, California," *Canadian Journal of Forest Research* 24 (1994): 21–31; Steven P. Norman, "A 500-Year Record of Fire from a Humid Coast Redwood Forest," *A Report to Save the Redwoods League* (U.S. Forest Service Redwood Sciences Laboratory, 2007); and for redwoods farther south, Gregory A. Jones and Will Russell, "Approximation of Fire-Return Intervals with Point Samples in the Southern Range of the Coast Redwood Forest, California, USA," *Fire Ecology* 11, no. 3, (2015): 80–94. I would be remiss not to include the classic by Emanuel Fritz, "The Role of Fire in the Redwood Region," *Journal of Forestry* 29 (1931): 939–50. For general background see John Evarts and Marjorie Popper, eds., *Coast Redwood: A Natural and Cultural History* (Cachuma Press, 2014), and still relevant Susan Schrepfer, *The Fight to Save the Redwoods: A History of Environmental Reform, 1917–1978* (Madison: University of Wisconsin Press, 1983).

4. For an excellent overview on burning the balds, see Stephen Underwood, Leonel Arguello, and Nelson Siefkin, "Restoring Ethnographic Landscapes and Natural Elements in Redwood National Park," *Ecological Restoration* 21, no. 4 (December 2003): 278–83.

5. Robert Boyd, ed., *Indians, Fire and the Land in the Pacific Northwest* (Corvallis: Oregon State University Press, 1999), especially Jeff LaLande and Reg Pullen, "Burning for a 'Fine and Beautiful Open Country': Native Uses of Fire in Southwestern Oregon," 255–76; Henry Lewis, *Patterns of Indian Burning in California: Ecology and Ethnohistory*, Ballena Press Anthropological Papers 1 (Ramona, Calif.: Ballena Press, 1973); M. Kat Anderson, *Tending the Wild: Native American Knowledge and the Management of California's Natural Resources* (Berkeley: University of California Press, 2005).

6. Taylor and Skinner, "Fire Regimes and Management," 204. On the dates for traditional burning, see Will Harling and Bill Tripp, "Western Klamath Restoration Partnership: A Plan for Restoring Fire Adapted Landscapes," submitted to Klamath National Forest (June 30, 2014).

7. My information derives from conversations, literature, a field trip, and a workshop organized by Ashland Forest Reserve in June 2016. I'm grateful to Darren Borgias for the invitation to participate, and to Shannon who joined us for a field trip to the East Antelope fire.

8. Cited in LaLande and Pullen, "Burning for a 'Fine and Beautiful Open Country,'" 255.

RESTORATION SINGS THE BLUES

1. Robert W. Mutch et al., *Forest Health in the Blue Mountains: A Management Strategy for Fire-Adapted Ecosystems*, General Technical Report PNW-

GTR-310, U.S. Forest Service, 1993, 1. For introducing me to Umatilla National Forest, I would like to thank David Powell and Chris Johnston, and for the Wallowa-Whitman National Forest, Noel Livingston, Steven Hawkins, and Larry Sandoval.

2.   Mutch et al., *Forest Health*, 12, 11, 13.

3.   Other basic documents include Boyd E. Wickman, "Forest Health in the Blue Mountains: The Influence of Insects and Diseases," in *Forest Health in the Blue Mountains: Science Perspectives*, General Technical Report PNW-GTR-295, ed. Thomas M. Quigley, U.S. Forest Service, 1992; William R. Gast Jr., et al., "Blue Mountains Forest Health Report: 'New Perspectives in Forest Health.'" Malheur, Umatilla, and Wallowa-Whitman National Forests (U.S. Forest Service, 1991); Ashley G. Juran, "Fire Regimes of Conifer Forests in the Blue Mountains," in Fire Effects Information System, U.S. Forest Service, Rocky Mountain Research Station, Missoula Fire Sciences Laboratory, accessed August 23, 2018, https://www.fs.fed.us/database/feis/fire_regimes/Blue_Mts_conifer/all.pdf.

4.   Nancy Langston, *Forest Dreams, Forest Nightmares: The Paradox of Old Growth in the Inland West* (Seattle: University of Washington Press, 1995), 46.

5.   Langston, 86.

6.   Langston, 163.

7.   Langston, 247. For a provocative take on how science and politics work on environmental issues, see Daniel Sarewitz, "How Science Makes Environmental Controversies Worse," *Environmental Science and Policy* 7 (2004): 385–403.

8.   Langston, 269, 304, 269.

9.   Langston, 295, 273.

10.   See David C. Powell, "Active Management of Dry Forests in the Blue Mountains: Silvicultural Considerations," White Paper F14-SO-WP-SILV-4, U.S. Forest Service, December 2014, for an argument in favor of mechanical treatments, which can be controlled, over prescribed fire, whose many interactions and aftershocks can't.

11.   Langston, *Forest Dreams*, 297.

12.   Langston, 290–300.

AN ECOLOGICAL AND SILVICULTURAL TOOL: HAROLD WEAVER

1.   Harold Weaver, "Fire and Its Relationship to Ponderosa Pine," in *Proceedings: 7th Tall Timbers Fire Ecology Conference* (Tallahassee, Fla.: Tall Timbers

Research Station, 1967), 127. I wish to thank Sonja Pyne for her help in locating newspaper references to Weaver in Oregon and Arizona.

2. Weaver, 127–28.

3. Weaver, 128.

4. Harold Weaver, "Ecological Changes in the Ponderosa Pine Forest of the Warm Springs Indian Reservation in Oregon," *Journal of Forestry* 57 (1959): 20.

5. Weaver, 7n.

6. Harold Weaver, "Fire as an Ecological and Silvicultural Factor in the Ponderosa Pine Region of the Pacific Slope," *Journal of Forestry* 41 (1943): 14–15.

7. Weaver, 15.

8. Weaver, 7.

9. Ben Avery, "Areas Burned to Cut Hazard, Help Growth," *Arizona Republic*, June 18, 1950, 13; "Forester Gets New Position in Washington," *Arizona Republic*, May 30, 1951, 11.

10. Harold H. Biswell et al., *Ponderosa Pine Management: A Task Force Evaluation of Controlled Burning in Ponderosa Pine Forests of Central Arizona*, Miscellaneous Publication 2 (Tallahassee, Fla.: Tall Timbers Research Station, 1973), 1–2.

EPILOGUE: THE PACIFIC NORTHWEST BETWEEN TWO FIRES

1. Donald P. Hanley, Jerry J. Kammenga, Chadwick D. Oliver, eds., *The Burning Decision: Regional Perspectives on Slash*, Institute of Forest Resources, Contribution 66, 1989, ix.

2. Isaac quote from "Preface," in Walstad et al., *Natural and Prescribed Fire*.

3. Agee, *Fire Ecology*, xi, 58.

4. The fires received wide attention. For the basics, see the Oregon Department of Forestry and Washington Department of Natural Resources websites, along with summaries in Northwest Fire Coordination Center reports. On the 2014 Carlton Complex, see also Michelle Nijhuis, "After One Record-Setting Wildfire, a Washington County Prepares For More," *High Country News*, August 3, 2015, http://www.hcn.org/issues/47.13/after-a-record-setting-wildfire-a-washington-county-prepares-for-the-next-one, and Methow Valley News, "Trial by Fire: The Methow Valley's Summer of Disaster," January 6, 2015, https://issuu.com/methowvalleypublishing/docs/2014trialbyfire_methowvalleynews/3. On the 2015 season see *Narrative Timeline of the Pacific Northwest 2015 Fire Season*, U.S. Forest Service, Pacific

Northwest Region, https://wfmrda.nwcg.gov/docs/_Reference_Materials /2015_Timeline_PNW_Season_FINAL.pdf. The Oregon Department of Forestry, Fire Protection Division, published a very useful *2015 Fire Season Report*, February 16, 2016, http://www.oregon.gov/ODF/Documents/Fire /2015_Protection_Division_Fire_Season_Report.pdf.

5. The Oregon Department of Forestry has an excellent summary of how it finances fire on its website. I found particularly helpful "History of Emergency Fire Cost Funding in Oregon" and accounts of the visits to Lloyd's to forestall the loss of fire insurance.

6. Information from site visit and the center's website: http://www.columbia breakswildfire.com.

7. Long quote from Eloise Hamilton, *Forty Years of Western Forestry* (Portland, Ore.: Western Forestry and Conservation Association, 1949), 3.

PROLOGUE: LAST FRONTIER, LOST FRONTIER

1. My Alaska trek was made possible by Ron Dunton, with assistance from the Joint Fire Science Program. I wish to thank Beth Ipsen and, in particular, Mike Roos for making my visit to the Alaska Fire Service productive. But there were many other people, not included in the specific acknowledgments elsewhere because their special expertise did not end up as a stand-alone topic. Let me thank them here: Thomas Kurth, Alison York, Peter Butteri, Jennifer Barnes, Doug Alexander, Brian Sorbel, Douglas Downs, Jay Wattenbarger, Randi Jandt, Bill Cramer, Larry Weddle, and Michael Butteri.

2. John McPhee, *Coming into the Country* (New York: Farrar, Straus, Giroux, 1977), 57.

3. Stephen Haycox, *Battleground Alaska: Fighting Federal Power in America's Last Wilderness* (Lawrence: University Press of Kansas, 2016), 24, 16.

THE ALASKAN PERSUASION

1. The concept of a "persuasion" comes from Marvin Meyers, *The Jacksonian Persuasion: Politics and Belief* (Stanford, Calif.: Stanford University Press, 1957). My development of the idea follows Stephen Haycox, especially *Battleground Alaska*.

2. My analysis follows Haycox, *Battleground Alaska*; see, especially, 21–24. I was myself introduced to Alaskan history in graduate school when a course in American West made Gruening's book required reading. This was in 1972,

just after the Alaska Native Claims Settlement Act, when Alaskan politics was a vital national topic. Gruening's text still looms over Alaskan studies, the Denali of Alaskan historiography.

3.   Quote from Haycox, *Battleground Alaska*, 13.

4.   Murry A. Taylor, *Jumping Fire: A Smokejumper's Memoir of Fighting Wildfire* (San Diego, Calif.: Harcourt, 2000), 219.

5.   Haycox, *Battleground Alaska*, 17. The ongoing fight over the Arctic National Wildlife Refuge (ANWR) has become largely symbolic. The real disruptor in Alaska's economy is fracking, not ANWR.

PYROPOLITICS, ALASKA STYLE

1.   A much richer account of the early years of Alaskan fire protection is available in Pyne, *Fire in America*, 497–512. That chronicle ends in the late 1970s. For the subsequent chronicle see Susan K. Todd and Holly Ann Jewkes, *Wildland Fire in Alaska: A History of Organized Fire Suppression and Management in the Last Frontier*, University of Alaska–Fairbanks, Agricultural and Forestry Experiment Station Bulletin No. 114, March 2006, and Mary Lynch, "Timeline," on file with AFS. The connection between the BLM's fire program in Alaska and its national agenda is reviewed in Stephen Pyne, *Between Two Fires: A Fire History of Contemporary America* (Tucson: University of Arizona Press, 2015), 74–84.

2.   Harold J. Lutz, *Aboriginal Man and White Man as Historical Causes of Fires in the Boreal Forest, with Particular Reference to Alaska*, Bulletin 65, Yale School of Forestry, Yale University, 1959, 23.

3.   On the USFS, see Lawrence W. Rakestraw, *A History of the United States Forest Service in Alaska* (Anchorage: Alaska Historical Commission, Department of Education, State of Alaska, 1981).

4.   Henry S. Graves, "The Forests of Alaska," *American Forests* 22 (1916): 33.

5.   General Land Office, Division of Forestry, *Field Handbook* (1940), title page. For a view of the situation by the man who would lead the AFCS, see W. J. McDonald, "Fire Under the Midnight Sun," *American Forests* 45 (April 1939): 168–69, 231.

6.   GLO, Division of Forestry, *Field Handbook*, title page. See also W. J. McDonald to Commissioner, General Land Office, "The Policy of Fire Control on the Public Domain of Alaska and Its Objectives," September 1, 1939, and Commissioner to McDonald, "Approval," January 10, 1949, Records of the Alaska Fire Control Service, Federal Records Center, Seattle.

7.   William R. Hunt, *Alaska* (New York: Norton, 1976), 115.

8. H. J. Lutz, *Ecological Effects of Forest Fires in the Interior of Alaska*, U.S. Department of Agriculture, Technical Bulletin No. 1133 (1956), 87, 89, 94; A. Starker Leopold and F. Fraser Darling, *Wildlife in Alaska: An Ecological Reconnaissance* (New York: Ronald Press, 1953), 55, 58–59.

9. C. E. Hardy, *Conflagration in Alaskan Forests, 1957*, unpublished report, Ogden, Utah, Intermountain Forest and Range Experiment Station, U.S. Forest Service, 1957, 2. See also James B. Craig, "Alaska Burns," *American Forests* 75 (October 1969): 1–3, 62–63, and John Clark Hunt, "Burning Alaska," *American Forests* 64 (August 1958): 12–15, 40–42.

10. On attitudes toward the 1947 and 1957 fires, see comments by Roger Robinson in Todd and Jewkes, *Wildland Fire in Alaska*, 25.

11. On the fire revolution and the BLM's role, see Pyne, *Between Two Fires*, 74–80.

12. Bureau of Land Management, "Position Paper: April 20, 1982," *Miscellaneous Background Material (October 81 to May 82)*. Position Papers, Directives, and Statistics, AFS files, June 9, 1982.

13. Alaska Department of Education, Fire Service Training Program, "Applicability of Wildlands Fire-Fighting Techniques for Structural Fires," Anchorage: Alaska Department of Education, 1977; quote from 79.

14. The most useful (and lively) account I've found is Roderick Nash, *Wilderness and the American Mind*, 4th ed. (New Haven, Conn.: Yale University Press, 2001), 272–315.

15. See Lynch, "Timeline." The best record of the details is the collection of *Miscellaneous Background Material (October 81 to May 82)—Position Papers, Directives, and Statistics—6/9/82*, held in the offices of the Alaska Fire Service in Fairbanks.

16. The quoted memos come from the files of the Alaska Fire Service, personally gathered by Mike Roos; a useful timeline was devised by Mary Lynch. I'm indebted to both of them. For quotes, see "Memorandum from BLM Director, BIFC to Director, Washington Office, March 30, 1981, Subject: Management of the Fire Program in Alaska."

17. *Alaska Interagency Fire Management Plan: Tanana/Minchumina Planning Area*, March 1982, 48–50.

18. "Position Paper: April 20, 1982."

LAST FRONTIER OF THE U.S. FOREST SERVICE

1. Langille quoted in Lawrence W. Rakestraw, *A History of the United States Forest Service in Alaska* (Anchorage: Alaska Historical Commission and U.S.

Forest Service, 1981), 35. I rely heavily on Rakestraw's detailed account for events into the 1970s.

2. R. S. Kellogg, "The Forests of Alaska," U.S. Forest Service, Bulletin 81 (Washington, D.C.: Government Printing Office, 1910), 22, 24.

3. Henry S. Graves, "The Forests of Alaska," *American Forestry*, 1916, 33. See also "The Alaskan Forests: An Interview with Henry S. Graves, the United States Forester," *Outlook* 112 (March 22, 1916): 679–80. John B. Guthrie, "Alaska's Interior Forests," *Journal of Forestry* 20 (1922): 363–73.

4. Rakestraw, *History*, 112. Hardy, *Conflagration*.

5. R. R. Robinson, "Forest and Range Fire Control in Alaska," *Journal of Forestry* 58 (1960): 448–53; Rakestraw, *History*, 148; Charles E. Hardy and James W. Franks, "Forest Fires in Alaska," Research Paper INT-5, U.S. Forest Service, 1963.

LIVE-FIRE ZONE

1. I want to thank Russ Long for his helpful introduction to the military zone's traits and ops. And Randi Jandt noted (personal communication, July 1, 2017) that "actually we have done some prescribed 'crown' fires, Alphabet Hills, 2004, was one done by BLM, 20,000 ac for wildlife habitat improvement; Farewell bison herd RX done by the State a few years ago and out by Delta for bison earlier this summer."

SPARKS OF IMAGINATION

1. Stuck quoted in Lutz, *Aboriginal Man*, 39–40.

2. Lutz, *Aboriginal Man*, 41–42.

3. Lutz, *Aboriginal Man*, 41. Statistics from Eric S. Kasischke, T. Scott Rupp, and David L. Verbyla, "Fire Trends in the Alaskan Boreal Forest," in *Alaska's Changing Boreal Forest*, ed. F. Stuart Chapin III et al. (New York: Oxford University Press, 2006), 285–301; quote on 291.

4. Kasischke et al., "Fire Trends in the Alaskan Boreal Forest," 291.

5. On Skyfire, see Kristine C. Harper, *Make It Rain: State Control of the Atmosphere in Twentieth-Century America* (Chicago, Ill.: University of Chicago Press, 2017), 168–75.

6. For an interesting survey of human interactions with fire in various ways, see La'ona DeWilde and F. Stuart Chapin III, "Human Impacts on the Fire Regime of Interior Alaska: Interactions Among Fuels, Ignition Sources, and Fire Suppression," *Ecosystems* 9, no. 8 (December 2006): 1342–53.

234 NOTES TO PAGES 196-203

IN THE BLACK

1. F. Stuart Chapin III et al., "Successional Processes in the Alaskan Boreal Forest," in *Alaska's Changing Boreal Forest*, ed. F. Stuart Chapin III et al. (New York: Oxford University Press, 2006), 113.

2. I have relied mostly on Chapin III et al., *Alaska's Changing Boreal Forest*, a marvelous compendium. Details of Holocene climate change can be found in the chapter by Andrea H. Lloyd et al., "Holocene Development of the Alaskan Boreal Forest," 62–78.

3. A handy summary of climate-related fire changes is available in "Wildfires," ACC-00100, Alaska Climate Change Adaptation Series, Alaska Center for Climate Assessment and Policy, University of Alaska–Fairbanks, http://www.snap.uaf.edu.

KENAI

1. I want to thank Kristi Bulock for setting a wonderful primer on Kenai fire. Others who contributed to the session, and beyond, are Andy Loranger, Doug Newbould, John Morton, Ed Berg, and Mike Hill. Randi Jandt also contributed some helpful observations. For a bibliography of scientific publications about the refuge, see https://www.fws.gov/refuge/Kenai/what_we _do/science/bibliography.html.

2. Fire history from E. E. Berg and R. S. Anderson, "Fire History of White and Lutz Spruce Forests on the Kenai Peninsula, Alaska, Over the Last Two Millennia as Determined from Soil Charcoal," *Forest Ecology and Management* 227, no. 3 (June 2006): 275–83; Starker Leopold and F. Fraser Darling, *Wildlife in Alaska: an Ecological Reconnaissance* (New York: Ronald Press, 1953), 58. For a useful summary, see "Fire Ecology and Regime Shift Due to Climate Change," Kenai National Wildlife Refuge, U.S. Fish and Wildlife Service, last updated September 26, 2012, https://www.fws.gov/refuge/Kenai /what_we_do/science/fire_ecology.html. On background fire information generally, see *Alaska Interagency Fire Management Plan: Kenai Peninsula Planning Area* (April 1984).

3. Leopold and Darling, *Wildlife in Alaska*, 58. The ubiquitous H. J. Lutz also investigated the question of moose history; see *History of the Early Occurrence of Moose on the Kenai Peninsula and in Other Sections of Alaska*, Alaska Forest Research Center, Miscellaneous Publication No. 1, June 1960.

4. See C. W. Slaughter, Richard J. Barney, and G. M. Hansen, eds., *Fire in the Northern Environment—A Symposium* (Portland, Ore.: Pacific Northwest Forest and Range Experiment Station, U.S. Forest Service, 1971).

5.   The bark beetle infestation has claimed the most national attention; for a journalistic account, see Andrew Nikiforuk, *Empire of the Beetle: How Human Folly and a Tiny Bug Are Killing North America's Great Forests* (Greystone Books, 2011), 4–30. A succinct summary of Project Skyfire is in Harper, *Make It Rain*, 168–75.

The drying peat has received special attention; see E. E. Berg et al., "Recent Woody Invasion of Wetland on the Kenai Peninsula Lowlands, South-Central Alaska: A Major Regime Shift After 18,000 Years of Wet *Sphagnum*-Sedge Peat Recruitment," *Canadian Journal of Forest Research* 39 (2009): 2033–46.

6.   Benjamin M. Jones, et al., "Presence of Rapidly Degrading Permafrost Plateaus in South-Central Alaska," *Crysophere* 10, no. 6 (November 2016): 2673–92.

7.   Conversations with Kenai staff, May 23, 2017; *Card Street Fire Fuels Treatment Effectiveness*, publication of Kenai NWR. John Morgan provided some extra details on the Caribou Hills and Card Street fires.

NORTH TO THE FUTURE: PLEISTOCENE TO PYROCENE

1.   Figures from Haycox, *Battleground Alaska*, 36.

2.   See, for example, John Imbrie and Katherine Palmer Imbrie, *Ice Ages: Solving the Mystery* (Cambridge, Mass.: Harvard University Press, 1986).

3.   See William Ruddiman, *Plows, Plagues, and Petroleum* (Princeton, N.J.: Princeton University Press, 2005).

EPILOGUE: ALASKA BETWEEN TWO FIRES

1.   F. Stuart Chapin III et al., "Increasing Wildfire in Alaska's Boreal Forest: Pathways to Potential Solutions of a Wicked Problem," *BioScience* 58, no 6 (June 2008): 531–40.

INDEX

ABOUT THE AUTHOR

Stephen J. Pyne is a Regents' Professor in the School of Life Sciences at Arizona State University and a former North Rim Longshot. Among his recent books are *Between Two Fires: A Fire History of Contemporary America* and To the Last Smoke, a series of regional fire surveys. He lives in Queen Creek, Arizona.